The Open University

C4

Completeness

This publication forms part of an Open University course. Details of this and other Open University courses can be obtained from the Student Registration and Enquiry Service, The Open University, PO Box 197, Milton Keynes, MK7 6BJ, United Kingdom: tel. +44 (0)870 333 4340, e-mail general-enquiries@open.ac.uk

Alternatively, you may visit the Open University website at http://www.open.ac.uk where you can learn more about the wide range of courses and packs offered at all levels by The Open University.

To purchase a selection of Open University course materials, visit the webshop at www.ouw.co.uk, or contact Open University Worldwide, Michael Young Building, Walton Hall, Milton Keynes, MK7 6AA, United Kingdom, for a brochure: tel. +44 (0)1908 858785, fax +44 (0)1908 858787, e-mail ouwenq@open.ac.uk

The Open University, Walton Hall, Milton Keynes, MK7 6AA.

First published 2006.

Copyright © 2006 The Open University

All rights reserved; no part of this publication may be reproduced, stored in a retrieval system, transmitted or utilised in any form or by any means, electronic, mechanical, photocopying, recording or otherwise, without written permission from the publisher or a licence from the Copyright Licensing Agency Ltd. Details of such licences (for reprographic reproduction) may be obtained from the Copyright Licensing Agency Ltd, 90 Tottenham Court Road, London W1T 4LP.

Open University course materials may also be made available in electronic formats for use by students of the University. All rights, including copyright and related rights and database rights, in electronic course materials and their contents are owned by or licensed to The Open University, or otherwise used by The Open University as permitted by applicable law.

In using electronic course materials and their contents you agree that your use will be solely for the purposes of following an Open University course of study or otherwise as licensed by The Open University or its assigns.

Except as permitted above you undertake not to copy, store in any medium (including electronic storage or use in a website), distribute, transmit or re-transmit, broadcast, modify or show in public such electronic materials in whole or in part without the prior written consent of The Open University or in accordance with the Copyright, Designs and Patents Act 1988.

Edited, designed and typeset by The Open University, using the Open University TeX System.

Printed and bound in the United Kingdom by The Charlesworth Group, Wakefield.

ISBN 0 7492 4137 3

1.1

Contents

Introduction	**4**
Study guide	4
1 Completeness	**5**
1.1 Cauchy sequences	5
1.2 Convergent Cauchy sequences	8
2 Complete metric spaces	**10**
2.1 Criteria for completeness	10
2.2 Examples of complete metric spaces	11
2.3 Completeness is not a topological invariant	16
3 The Contraction Mapping Theorem	**19**
3.1 Fixed-points and the Contraction Mapping Theorem	19
3.2 Applications of the Contraction Mapping Theorem	26
4 Completion	**30**
4.1 Defining the completion	30
4.2 Constructing the completion of a space	31
Solutions to problems	**37**
Index	**40**

Introduction

In the previous unit we studied sequences in topological spaces and saw that, for metric spaces, compactness can be described entirely in terms of the existence of convergent sequences. We also identified some compact subsets of $C[0,1]$ for its usual topology.

We now continue our investigation of sequences by looking more closely at the role they play in metric spaces. We are motivated by the following question.

When can we know that a given sequence converges without first having to find its limit?

We first explain what we mean by a sequence that 'appears to converge'. We then identify those metric spaces for which such sequences *do* converge — we call these *complete* metric spaces.

We then discuss the *Contraction Mapping Theorem* — a powerful theorem that gives sufficient conditions to ensure that a continuous mapping of a certain type on a complete metric space must have a fixed-point. The Contraction Mapping Theorem is a surprisingly useful result and we give several examples of its applications. In *Unit C5, Fractals*, you will see that the Contraction Mapping Theorem plays an important role in the investigation of fractals.

Finally, we show how to extend any metric space to a complete metric space. This provides a natural way to embed an arbitrary metric space into a space for which the Contraction Mapping Theorem may be used.

Study guide

In Section 1, *Completeness*, we investigate properties of sequences in $\mathbb{R}$ that guarantee convergence for $d^{(1)}$. We define the key concept underlying the definition of completeness — that of a *Cauchy sequence* — and establish some basic properties of Cauchy sequences.

In Section 2, *Complete metric spaces*, we give examples of complete metric spaces and develop some basic theory of such spaces. You should study this section closely as the results are needed throughout the unit.

In Section 3, *The Contraction Mapping Theorem*, we present an important result that illustrates the usefulness of completeness and discuss some of its applications. This is the longest section in the unit and it contains many problems; we suggest that you spread this work over two study sessions.

In Section 4, *Completion*, we discuss how to extend any metric space to a complete metric space. If you are short of time, concentrate on Subsection 4.1, in which we explain how to find the 'smallest' complete metric spaces containing many of our standard metric spaces. Subsection 4.2 is not assessed.

There is no software associated with this unit.

1 Completeness

After working through this section, you should be able to:
▶ determine whether a given sequence is a *Cauchy sequence*;
▶ describe some properties of Cauchy sequences;
▶ define a *complete* metric space.

In analysis, we are frequently given a sequence in a metric space, and we wish to know whether it is convergent.

In *Unit C3*, we defined a convergent sequence as follows.

Definition

Let (X, d) be a metric space. A sequence (a_n) in X **converges** to $a \in X$ if $(d(a_n, a))$ is a null sequence.

When using this definition to show that a given sequence is convergent, we must first identify a possible limit of the sequence and show that the sequence converges to it. In many instances, however, it is difficult to identify a possible limit. For example, consider the sequence of real numbers (a_n) given by

$$a_n = 1 + \frac{1}{2^2} + \frac{1}{3^2} + \cdots + \frac{1}{n^2}.$$

The limit of this sequence is not obvious. Fortunately, we have the Monotone Convergence Theorem, which states that a non-decreasing sequence of real numbers that is bounded above is convergent. This sequence is non-decreasing (since $a_{n+1} - a_n = \frac{1}{(n+1)^2} > 0$) and is bounded above by 2 (since $a_n \leq 1 + \int_1^n \frac{1}{x^2}\,dx$). So, by the Monotone Convergence Theorem, we know that (a_n) converges even though we cannot identify its limit.

The limit of this sequence is in fact $\pi^2/6$ — this was first calculated in the eighteenth century by Leonhard Euler.

Our next objective is to identify a property which ensures that, for a large class of metric spaces, sequences with this property are convergent. This enables us to state that many sequences are convergent without our having to identify their limits. Unfortunately, we cannot apply a theorem similar to the Monotone Convergence Theorem to a large class of metric spaces, since the elements of many metric spaces have no natural ordering.

1.1 Cauchy sequences

The following result gives a property that every convergent sequence possesses — the terms of a convergent sequence (a_n) get arbitrarily close to each other when n is sufficiently large.

Lemma 1.1

Let (X, d) be a metric space and let (a_n) be a convergent sequence in X. Then, for each $\varepsilon > 0$, there is an $N \in \mathbb{N}$ such that

$d(a_n, a_m) < \varepsilon$ for all $n, m > N$.

Proof Let (a_n) be a convergent sequence in X. Then there is an $a \in X$ such that
$$a_n \xrightarrow{d} a \text{ as } n \to \infty.$$
Let $\varepsilon > 0$. Then there is an $N \in \mathbb{N}$ such that, for all $n > N$,
$$d(a_n, a) < \tfrac{1}{2}\varepsilon.$$
We deduce from the Triangle Inequality that, for all $n, m > N$,
$$d(a_n, a_m) \leq d(a_n, a) + d(a, a_m) < \tfrac{1}{2}\varepsilon + \tfrac{1}{2}\varepsilon = \varepsilon,$$
as required. ∎

Recall that $a_n \xrightarrow{d} a$ if and only if $d(a_n, a) \to 0$.

The property of convergent sequences given in Lemma 1.1 is the one we seek.

> **Definition**
>
> Let (X, d) be a metric space and let (a_n) be a sequence in X. Then (a_n) is a **Cauchy sequence** if, for each $\varepsilon > 0$, there is an $N \in \mathbb{N}$ such that
> $$d(a_n, a_m) < \varepsilon \quad \text{for all } n, m > N.$$

Augustin-Louis Cauchy (1789–1857) was a prolific French mathematician whose name occurs in many areas of analysis. His book *Cours d'analyse* was one of the first to place the foundations of calculus on a rigorous basis.

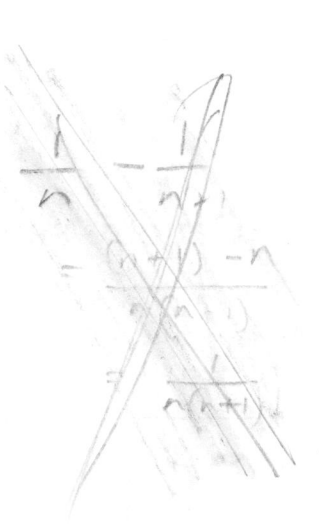

Remarks

(i) Lemma 1.1 states that any convergent sequence in a metric space is a Cauchy sequence.

(ii) This definition does not require us to identify a limit of the sequence.

(iii) We may assume that $m > n$ in the definition since if $n = m$, then $d(a_n, a_m) = 0$, which is certainly smaller than ε; and if $m < n$, we can interchange n and m since $d(a_n, a_m) = d(a_m, a_n)$.

(iv) To show that a given sequence is *not* a Cauchy sequence, it is enough to find just *one* $\varepsilon > 0$ such that, for each $N \in \mathbb{N}$, there are $n, m > N$ with $d(a_n, a_m) \geq \varepsilon$.

(v) If we wish to emphasize the metric, we say that (a_n) is a Cauchy sequence *for d*, or *with respect to d*, or that (a_n) is a d-Cauchy sequence.

Problem 1.1

Determine whether each of the following sequences is a $d^{(1)}$-Cauchy sequence in $\mathbb{R}$.

(a) $(1/n)$ (b) (n)

We now derive some basic properties of Cauchy sequences.

> **Lemma 1.2**
>
> Let (X, d) be a metric space, and let (a_n) be a Cauchy sequence in X. Then (a_n) lies in an open ball of X — that is, there are an $a \in X$ and an $M > 0$ such that
> $$\{a_n : n \in \mathbb{N}\} \subseteq B_d(a, M).$$

Proof Since (a_n) is a Cauchy sequence, there is an $N \in \mathbb{N}$ such that

$$d(a_n, a_m) < 1 \quad \text{for all } n, m > N.$$

Let $a = a_{N+1}$; then

$$d(a_n, a) < 1 \quad \text{whenever } n > N.$$

So, for each $n \in \mathbb{N}$,

$$d(a_n, a) \leq \max\{1, d(a_1, a), d(a_2, a), d(a_3, a), \ldots, d(a_N, a)\}.$$

This implies that

$$\{a_n : n \in \mathbb{N}\} \subseteq B_d(a, M),$$

where $M = 1 + \max\{1, d(a_1, a), d(a_2, a), d(a_3, a), \ldots, d(a_N, a)\}$. ∎

> **Corollary 1.3**
>
> Each $d^{(1)}$-Cauchy sequence in $\mathbb{R}$ is bounded.

Proof Let (a_n) be a $d^{(1)}$-Cauchy sequence in $\mathbb{R}$. By Lemma 1.2, there are an $a \in \mathbb{R}$ and an $M > 0$ such that

$$|a_n - a| < M \quad \text{for each } n \in \mathbb{N}.$$

By the Triangle Inequality,

$$|a_n| = |a_n - a + a| \leq |a_n - a| + |a| < M + |a|,$$

for each $n \in \mathbb{N}$. So (a_n) is bounded. ∎

We now show that if a subsequence of a Cauchy sequence converges, then the whole sequence must converge. This is not true for sequences in general — for example, the sequence $(1, -1, 1, -1, \ldots)$ does not converge but has the convergent subsequence $(1, 1, 1, \ldots)$.

> **Lemma 1.4**
>
> Let (X, d) be a metric space and let (a_n) be a Cauchy sequence in X. If (a_n) has a convergent subsequence with limit a in X, then (a_n) is convergent with limit a.

Proof Let (a_{n_k}) be a subsequence of (a_n) that converges to $a \in X$. We prove that (a_n) also converges to a.

Let $\varepsilon > 0$ be given. Since $a_{n_k} \to a$ as $k \to \infty$, there is a $K \in \mathbb{N}$ such that, for all $k > K$,

$$d(a_{n_k}, a) < \tfrac{1}{2}\varepsilon.$$

Since (a_n) is a Cauchy sequence there is an $N \in \mathbb{N}$ such that

$$d(a_n, a_m) < \tfrac{1}{2}\varepsilon \quad \text{for all } n, m > N.$$

Let $k > K$ be chosen so that $n_k > N$; then

$$d(a_n, a_{n_k}) < \tfrac{1}{2}\varepsilon \quad \text{for all } n > N.$$

Moreover, since $k > K$,

$$d(a_{n_k}, a) < \tfrac{1}{2}\varepsilon.$$

By the Triangle Inequality,

$$d(a_n, a) \leq d(a_n, a_{n_k}) + d(a_{n_k}, a) < \tfrac{1}{2}\varepsilon + \tfrac{1}{2}\varepsilon = \varepsilon,$$

for all $n > N$.

Since $\varepsilon > 0$ is arbitrary, (a_n) converges to a in (X, d). ∎

1.2 Convergent Cauchy sequences

We have seen that in any metric space, each convergent sequence is a Cauchy sequence. We now show that in $\mathbb{R}$ (with its usual metric) the Cauchy sequences are precisely the convergent sequences, thus providing further evidence that the Cauchy property is the one that we sought. We already know from Lemma 1.1 that every convergent sequence in $\mathbb{R}$ is a Cauchy sequence, and we must now show that each Cauchy sequence is convergent in $\mathbb{R}$.

Lemma 1.1.

In order to prove this, we recall from Corollary 1.3 that a Cauchy sequence in $\mathbb{R}$ is bounded, and then construct a monotonic subsequence of the Cauchy sequence. By the Monotone Convergence Theorem this subsequence is convergent, and so by Lemma 1.4 the original Cauchy sequence is convergent. We need the following lemma.

Lemma 1.5

Let (a_n) be a sequence of real numbers. Then (a_n) contains a monotonic subsequence — that is, there is a subsequence (a_{n_k}) such that:

either

$$a_{n_1} \leq a_{n_2} \leq a_{n_3} \leq \cdots \leq a_{n_k} \leq \cdots$$

or

$$a_{n_1} \geq a_{n_2} \geq a_{n_3} \geq \cdots \geq a_{n_k} \geq \cdots.$$

The lemma does not state that the monotonic subsequence converges. For example, the sequence $(1, -1, 2, -2, 3, -3, 4, -4, \ldots)$ contains the strictly increasing subsequence $(1, 2, 3, \ldots)$, which does not converge.

Proof Let (a_n) be a sequence of real numbers. We say that a_n is a *peak term* of the sequence if no later term of the sequence is larger than a_n — that is, a_n is a peak term if $a_m \leq a_n$ for each $m > n$.

If there are infinitely many peak terms, let n_k be chosen so that a_{n_k} is the kth peak term. By the definition of a peak term, for each $k \in \mathbb{N}$, $a_m \leq a_{n_k}$ whenever $m > n_k$; in particular, $a_{n_{k+1}} \leq a_{n_k}$. Thus (a_{n_k}) is a decreasing sequence.

If there are only finitely many peak terms, then there is an $n_1 \in \mathbb{N}$ such that all the peak terms occur before n_1. In particular, a_{n_1} is not a peak term and so there must be an $n_2 > n_1$ such that $a_{n_2} > a_{n_1}$. Since a_{n_2} is not a peak term, we can find an $n_3 > n_2$ such that $a_{n_3} > a_{n_2}$. This process can be continued indefinitely, resulting in an increasing subsequence (a_{n_k}).

In each case, there exists a monotonic subsequence of (a_n). ∎

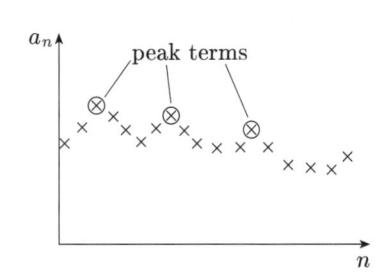

Figure 1.1

We can now show that every Cauchy sequence in $\mathbb{R}$ is convergent.

> **Theorem 1.6**
>
> Every $d^{(1)}$-Cauchy sequence in $\mathbb{R}$ is $d^{(1)}$-convergent.

Proof

Let (a_n) be a Cauchy sequence of real numbers.

By Corollary 1.3, there is a positive real number M such that

$|a_n| < M$ for each $n \in \mathbb{N}$.

By Lemma 1.5, there exists a monotonic subsequence (a_{n_k}) of (a_n). Since (a_n) is a bounded sequence, the monotonic subsequence is also bounded. Hence by the Monotone Convergence Theorem, the subsequence is convergent. Thus by Lemma 1.4, (a_n) is convergent. ∎

Combining this result with our earlier observation that convergent sequences in metric spaces are Cauchy sequences, we obtain the following result.

> **Corollary 1.7**
>
> A sequence in $\mathbb{R}$ is $d^{(1)}$-convergent if and only if it is a $d^{(1)}$-Cauchy sequence.

We have shown that the $d^{(1)}$-Cauchy sequences in $\mathbb{R}$ are precisely the $d^{(1)}$-convergent sequences. This enables us to determine whether a sequence in $\mathbb{R}$ is convergent without needing to identify the limit of the sequence — exactly what we wanted.

We now investigate other metric spaces for which this is true. We know that for any metric space, each convergent sequence is a Cauchy sequence, and so we are interested in those metric spaces for which the converse also holds — that is, metric spaces for which each Cauchy sequence is convergent.

> **Definition**
>
> A metric space (X, d) is **complete** if each Cauchy sequence in X is convergent; otherwise it is **incomplete**.

Theorem 1.6 tells us that $(\mathbb{R}, d^{(1)})$ is a complete metric space.

Problem 1.2

Show that $((0, 1], d^{(1)})$ is incomplete.

In the next section, we show that many of our standard examples of metric spaces are complete, and we develop some methods for identifying complete metric spaces.

2 Complete metric spaces

After working through this section, you should be able to:
- ▶ recognize several standard complete metric spaces;
- ▶ state and use some sufficient conditions for a metric space to be complete;
- ▶ explain why completeness is not a topological invariant;
- ▶ determine whether two given metric spaces are *isometric*.

So far, our only example of a complete metric space is $(\mathbb{R}, d^{(1)})$. We now develop criteria that guarantee that a given space is complete and identify some other complete metric spaces.

2.1 Criteria for completeness

We have seen that $(\mathbb{R}, d^{(1)})$ is complete, but the metric space $((0, 1], d^{(1)})$ is incomplete since there are Cauchy sequences in $(0, 1]$ that converge to 0 in $\mathbb{R}$. The following result shows that $(A, d_A^{(1)})$ is complete if A is any *closed* subset of $\mathbb{R}$. The proof relies on the characterization of the closure of a set in terms of sequences: if (X, d) is a metric space and $A \subseteq X$, then $a \in \mathrm{Cl}(A)$ if and only if there is a sequence in A that converges to a.

Unit C3, Theorem 2.2.

Proposition 2.1

Let (X, d) be a metric space and let $A \subseteq X$. Let d_A denote the induced metric on A from d.

(a) If (A, d_A) is a complete metric space, then A is closed.

(b) If (X, d) is a complete metric space and A is closed, then (A, d_A) is a complete metric space.

Proof

(a) *Let (A, d_A) be complete. We show that A is closed.*

Since a set is closed if it coincides with its closure, and $A \subseteq \mathrm{Cl}(A)$, it is enough to show that $\mathrm{Cl}(A) \subseteq A$. So let $a \in \mathrm{Cl}(A)$ and let (a_n) be a sequence in A that converges to a in X. By Lemma 1.1, (a_n) is a d-Cauchy sequence in X, and so it is a d_A-Cauchy sequence in A. But (A, d_A) is complete, and so (a_n) converges to a point in A. Since limits are unique in a metric space, this limit is a. Thus $a \in A$, so $\mathrm{Cl}(A) \subseteq A$.

Unit C3, Proposition 2.1.

(b) *Let (X, d) be complete and let $A \subseteq X$ be closed. We show that (A, d_A) is complete.*

If (a_n) is a d_A-Cauchy sequence in A, then it is also a d-Cauchy sequence in X. Since X is complete, (a_n) converges to some $a \in X$. Since (a_n) is in A, the limit a belongs to $\mathrm{Cl}(A)$. Since A is closed, $a \in A$, and so (a_n) is d_A-convergent in A. Thus (A, d_A) is complete. ∎

It follows from Proposition 2.1 that if $A \subseteq \mathbb{R}$, then $(A, d^{(1)})$ is complete if and only if A is closed.

Problem 2.1

For which of the following sets A is $(A, d_A^{(1)})$ complete?
(a) $A = [0, 1]$ (b) $A = [0, 1)$ (c) $A = (0, 1)$

Proposition 2.1 implies that compact subsets of complete metric spaces are complete, since they are closed. We can say more by using the characterization of compact metric spaces in terms of sequences: (X, d) is a compact metric space if and only if every sequence in X has a convergent subsequence.

Unit C2, Theorem 3.8.

Unit C3, Theorems 4.2 and 4.3.

Theorem 2.2

Every compact metric space is complete.

Proof Let (X, d) be a compact metric space and suppose that (a_n) is a Cauchy sequence in X. Since X is compact, (a_n) has a convergent subsequence, and Lemma 1.4 implies that (a_n) is convergent. ∎

Problem 2.2

Let (X, d) be a compact metric space, let (Y, e) be a metric space, and let $f: X \to Y$ be a continuous onto function. Is (Y, e) complete? What can you deduce if f is not onto?

2.2 Examples of complete metric spaces

Our first example involves the discrete metric. Recall that

$$d_0(x, y) = \begin{cases} 0 & \text{if } x = y, \\ 1 & \text{if } x \neq y. \end{cases}$$

Problem 2.3

Let X be a non-empty set. Show that each d_0-Cauchy sequence in X is eventually constant.

It follows from Problem 2.3 that if X is any non-empty set, then any d_0-Cauchy sequence in X is d_0-convergent and so (X, d_0) is a complete metric space.

Unit C3, Worked problem 1.2.

We now ask you to show that a *finite* metric space is complete.

Problem 2.4

Let $X = \{x_1, x_2, \ldots, x_n\}$ be a finite set and let d be a metric on X. Prove that (X, d) is complete.

Hint Consider the minimum of all the possible values of $d(x_i, x_j)$, for $i \neq j$.

The above examples are particularly simple metric spaces. Our remaining examples of complete metric spaces are more interesting.

The completeness of $(\mathbb{R}^k, d^{(k)})$

Recall from Corollary 1.7 that $(\mathbb{R}, d^{(1)})$ is a complete metric space. We now show that $(\mathbb{R}^k, d^{(k)})$ is a complete metric space, for any $k \geq 1$.

In *Unit C3*, we saw that a sequence $(\mathbf{a}_n)$ in $\mathbb{R}^k$ converges to $\mathbf{a}$ with respect to the Euclidean metric if and only if each component sequence (a_n^j) converges to a^j, for $1 \leq j \leq k$. Using the same idea, we can show that $(\mathbb{R}^k, d^{(k)})$ is complete — that is, each Cauchy sequence in $\mathbb{R}^k$ converges to a point in $\mathbb{R}^k$.

Unit C3, Theorem 1.7.

Here $\mathbf{a} = (a^1, a^2, \ldots, a^k)$ and $\mathbf{a}_n = (a_n^1, a_n^2, \ldots, a_n^k)$.

We first consider the relationship between a Cauchy sequence in $\mathbb{R}^k$ and the corresponding component sequences.

Lemma 2.3

A sequence $(\mathbf{a}_n)$ is a $d^{(k)}$-Cauchy sequence in $\mathbb{R}^k$ if and only if each component sequence (a_n^j) is a $d^{(1)}$-Cauchy sequence in $\mathbb{R}$, for $1 \leq j \leq k$.

Proof Let $(\mathbf{a}_n)$ be a $d^{(k)}$-Cauchy sequence in $\mathbb{R}^k$. Given $\varepsilon > 0$, choose an $N \in \mathbb{N}$ such that
$$d^{(k)}(\mathbf{a}_n, \mathbf{a}_m) < \varepsilon \quad \text{for all } n, m > N.$$
For each $j \in \{1, 2, \ldots, k\}$,
$$|a_n^j - a_m^j| \leq d^{(k)}(\mathbf{a}_n, \mathbf{a}_m) < \varepsilon \quad \text{for all } n, m > N,$$
and so each (a_n^j) is a $d^{(1)}$-Cauchy sequence.

Conversely, suppose that (a_n^j) is a $d^{(1)}$-Cauchy sequence in $\mathbb{R}$, for each $j \in \{1, 2, \ldots, k\}$. Given $\varepsilon > 0$ and $j \in \{1, 2, \ldots, k\}$, choose an $N_j \in \mathbb{N}$ such that
$$|a_n^j - a_m^j| < \varepsilon/\sqrt{k} \quad \text{for all } n, m > N_j.$$
Then, for $n, m > \max\{N_1, N_2, \ldots, N_k\}$,
$$d^{(k)}(\mathbf{a}_n, \mathbf{a}_m)^2 = \sum_{j=1}^{k} |a_n^j - a_m^j|^2 < \sum_{j=1}^{k} (\varepsilon/\sqrt{k})^2 = \varepsilon^2,$$
and so $d^{(k)}(\mathbf{a}_n, \mathbf{a}_m) < \varepsilon$. Thus $(\mathbf{a}_n)$ is a $d^{(k)}$-Cauchy sequence in $\mathbb{R}^k$. ∎

Lemma 2.3 enables us to show that $(\mathbb{R}^k, d^{(k)})$ is complete.

Corollary 2.4

The space $(\mathbb{R}^k, d^{(k)})$ is complete.

Proof Let $(\mathbf{a}_n)$ be a $d^{(k)}$-Cauchy sequence in $\mathbb{R}^k$. By Lemma 2.3, (a_n^j) is a $d^{(1)}$-Cauchy sequence in $\mathbb{R}$, for $1 \leq j \leq k$, and so by Corollary 1.7 it converges to a^j, say. Thus $(\mathbf{a}_n)$ converges to $\mathbf{a} = (a^1, a^2, \ldots, a^k)$. Hence $(\mathbb{R}^k, d^{(k)})$ is complete. ∎

Unit C3, Theorem 1.7.

Complete spaces of functions

We now turn to an important example of a complete metric space that is not a subset of a Euclidean space: $(C[0,1], d_{\max})$. We have already studied this space in some detail. Recall that $d_{\max}$ is given by

$$d_{\max}(f,g) = \max\{|g(x) - f(x)| : x \in [0,1]\},$$

and measures the maximum distance between two continuous functions on $[0,1]$.

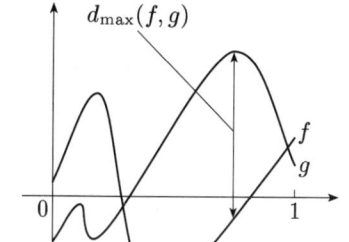

Unit A2, Subsection 2.3.

In this subsection, we prove the following result.

> **Theorem 2.5**
>
> The space $(C[0,1], d_{\max})$ is complete.

Figure 2.1

The proof of this result is quite sophisticated, and in order to prove it we use the notion of uniform convergence, which we introduced in *Unit C3*.

Recall that uniform convergence for a sequence of functions from $[0,1]$ to $\mathbb{R}$ is defined as follows.

Unit C3, Subsection 3.2.

> **Definition**
>
> A sequence (f_n) of functions $f_n: [0,1] \to \mathbb{R}$ **converges uniformly** on $[0,1]$ to the function $f: [0,1] \to \mathbb{R}$ if there is an $N \in \mathbb{N} \cup \{0\}$ such that:
>
> (a) the function $f_n - f$ is bounded for each $n > N$;
>
> (b) the sequence (M_n) defined by
>
> $$M_n = \sup\{|f_{n+N}(x) - f(x)| : x \in [0,1]\} \quad \text{for all } n \geq 1$$
>
> is a null sequence.

Proof of Theorem 2.5 Suppose that we are given a Cauchy sequence of functions (f_n) in $C[0,1]$.

You may wish to omit this on a first reading.

(a) We first identify a possible limit f of the sequence (f_n). We do this by observing that, for each $x \in [0,1]$, $(f_n(x))$ is a $d^{(1)}$-Cauchy sequence in $\mathbb{R}$. Since $(\mathbb{R}, d^{(1)})$ is complete, this means that we can define $f: [0,1] \to \mathbb{R}$ by $f(x) = \lim_{n \to \infty} f_n(x)$.

(b) We show that (f_n) converges uniformly to f.

(c) We deduce that f is continuous and that $d_{\max}(f_n, f) \to 0$.

Thus (f_n) is $d_{\max}$-convergent to f in $C[0,1]$, and so $(C[0,1], d_{\max})$ is complete.

(a) *Identify a possible limit f of the sequence (f_n)*

For each $x \in [0,1]$, and for all $n, m \in \mathbb{N}$,

$$|f_n(x) - f_m(x)| = d^{(1)}(f_n(x), f_m(x)) \leq d_{\max}(f_n, f_m).$$

Thus for each $x \in [0,1]$, $(f_n(x))$ is a $d^{(1)}$-Cauchy sequence in $\mathbb{R}$. By Theorem 1.6, $(f_n(x))$ is $d^{(1)}$-convergent, and we can define $f: [0,1] \to \mathbb{R}$ by

$$f(x) = \lim_{n \to \infty} f_n(x).$$

This function f is our proposed limit of the sequence (f_n).

(b) *Show that (f_n) converges uniformly to f*

We first show that the function $f_n - f$ is bounded for all $n \in \mathbb{N}$. To do this, we observe that Lemma 1.2 implies that there are an $M > 0$ and a $g \in C[0,1]$ such that

$$f_n \in B_{d_{\max}}(g, M) \quad \text{for each } n \in \mathbb{N}.$$

Hence for each $x \in [0,1]$ and each $n \in \mathbb{N}$,

$$|f_n(x) - g(x)| \leq d_{\max}(f_n, g) < M.$$

The Triangle Inequality now implies that, for each $x \in [0,1]$ and each $n \in \mathbb{N}$,

$$|f(x) - g(x)| \leq |f(x) - f_n(x)| + |f_n(x) - g(x)| < |f(x) - f_n(x)| + M.$$

For each $x \in [0,1]$, $|f(x) - f_n(x)| \to 0$ as $n \to \infty$, and so

$$|f(x) - g(x)| \leq M.$$

The Triangle Inequality now implies that, for each $x \in [0,1]$ and each $n \in \mathbb{N}$,

$$|f(x) - f_n(x)| \leq |f(x) - g(x)| + |g(x) - f_n(x)| < 2M.$$

Thus the function $f_n - f$ is bounded, for each $n \in \mathbb{N}$.

We now show that

$$M_n = \sup\{|f_n(x) - f(x)| : x \in [0,1]\}$$

is a null sequence — this implies that (f_n) converges uniformly to f.

Let $\varepsilon > 0$. Since (f_n) is a $d_{\max}$-Cauchy sequence, there is an $N \in \mathbb{N}$ such that, for all $m, n > N$,

$$d_{\max}(f_n, f_m) < \tfrac{1}{2}\varepsilon.$$

Suppose that $x \in [0,1]$. Since $(f_n(x))$ converges to $f(x)$, there is an $M \in \mathbb{N}$ such that, for all $m > M$,

$$|f(x) - f_m(x)| < \tfrac{1}{2}\varepsilon.$$

Let $n > N$ and $m > \max\{M, N\}$. By the Triangle Inequality,

$$\begin{aligned}|f(x) - f_n(x)| &\leq |f(x) - f_m(x)| + |f_m(x) - f_n(x)| \\ &\leq |f(x) - f_m(x)| + d_{\max}(f_n, f_m) \\ &< \tfrac{1}{2}\varepsilon + \tfrac{1}{2}\varepsilon = \varepsilon.\end{aligned}$$

Note the introduction of the function f_m and the consequent disappearance of the dependence on x in the bound for $|f(x) - f_n(x)|$. This is a *uniformization argument*: we find an estimate of $|f(x) - f_n(x)|$ that does not depend on the choice of x. This is a useful technique.

Since $x \in [0,1]$ is arbitrary, we deduce that, for $n > N$,

$$M_n = \sup\{|f(x) - f_n(x)| : x \in [0,1]\} \leq \varepsilon.$$

Since $\varepsilon > 0$ is arbitrary, (M_n) is a null sequence. Thus (f_n) converges uniformly to f.

(c) *Deduce that $f \in C[0,1]$ and that $d_{max}(f_n, f) \to 0$ as $n \to \infty$*

Since (f_n) converges uniformly to f, f is continuous on $[0,1]$, and so is an element of $C[0,1]$, and $d_{\max}(f_n, f) \to 0$ as $n \to \infty$.

Unit C3, Theorem 3.3.

Hence (f_n) is $d_{\max}$-convergent to f in $C[0,1]$, proving the theorem. ∎

Example 2.1

In the Exercises for *Unit A2*, we gave another example of a metric space of functions — namely, $(C^1[0,1], d)$, where

$$C^1[0,1] = \{f \in C[0,1] : f \text{ is differentiable and } f' \in C[0,1]\}$$

and

$$d(f,g) = d_{\max}(f,g) + d_{\max}(f',g').$$

Exercise A2.7.

Here, f' is the derivative of f.

It can be shown that this is also a complete metric space. The proof of this uses the fact that $(C[0,1], d_{\max})$ is complete, together with some careful analysis of the convergence of sequences of derivative functions.

We do not give the proof here.

In the next example, we give a metric on $C[0,1]$ for which the resulting space is *not* complete.

Example 2.2 $C[0,1]$ *with the integration metric*

The integration metric on $C[0,1]$ is given by

$$d(f,g) = \int_0^1 |g(x) - f(x)|\, dx.$$

We saw in Exercise *A2.8* that this defines a metric on $C[0,1]$.

Define $f_n \in C[0,1]$ by

$$f_n(x) = \begin{cases} 0 & \text{if } 0 \leq x \leq \tfrac{1}{2}\left(1 - \tfrac{1}{n}\right), \\ 2n\left(x - \tfrac{1}{2}\left(1 - \tfrac{1}{n}\right)\right) & \text{if } \tfrac{1}{2}\left(1 - \tfrac{1}{n}\right) < x < \tfrac{1}{2}, \\ 1 & \text{if } \tfrac{1}{2} \leq x \leq 1. \end{cases}$$

Either by integrating, or by finding the areas of appropriate triangles, we can easily show that if $N \in \mathbb{N}$ and $m > n > N$, then

$$d(f_m, f_n) = \frac{1}{4}\left(\frac{1}{n} - \frac{1}{m}\right) < \frac{1}{2N}.$$

Hence (f_n) is a d-Cauchy sequence in $C[0,1]$.

Now define $f: [0,1] \to \mathbb{R}$ by

$$f(x) = \begin{cases} 0 & \text{if } 0 \leq x < \tfrac{1}{2}, \\ 1 & \text{if } \tfrac{1}{2} \leq x \leq 1. \end{cases}$$

Then

$$\int_0^1 |f(x) - f_n(x)|\, dx = \int_{\frac{1}{2}(1-\frac{1}{n})}^{\frac{1}{2}} 2n\left(x - \frac{1}{2}\left(1 - \frac{1}{n}\right)\right) dx$$

$$= \left[n\left(x - \frac{1}{2}\left(1 - \frac{1}{n}\right)\right)^2\right]_{\frac{1}{2}(1-\frac{1}{n})}^{\frac{1}{2}}$$

$$= n\left(\frac{1}{2n}\right)^2 = \frac{1}{4n}.$$

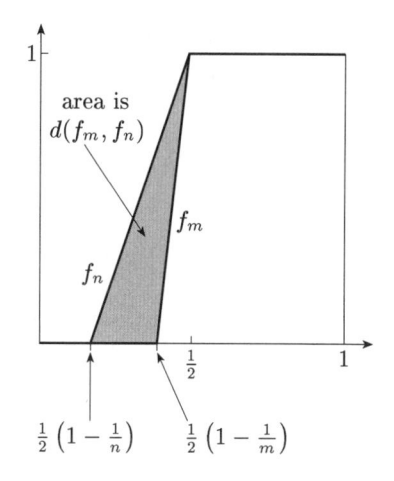

Figure 2.2

Thus f is a candidate for the limit of the sequence (f_n). However, f is *not* an element of $C[0,1]$, since it is not continuous.

In fact, for *no* function $g \in C[0,1]$ does $d(f_n, g) \to 0$ as $n \to \infty$. To see why, it is enough to show that if $g \in C[0,1]$, then

$$\int_0^1 |g(x) - f(x)|\, dx > 0.$$

For then we can use the Reverse Triangle Inequality to estimate

$$d(f_n, g) = \int_0^1 |g(x) - f_n(x)|\, dx$$
$$\geq \int_0^1 |g(x) - f(x)|\, dx - \int_0^1 |f(x) - f_n(x)|\, dx$$
$$= \int_0^1 |g(x) - f(x)|\, dx - \frac{1}{4n}.$$

So if n is large enough, then

$$d(f_n, g) > \tfrac{1}{2} \int_0^1 |g(x) - f(x)|\, dx > 0,$$

and so (f_n) does not converge to g.

We now estimate $\int_0^1 |g(x) - f(x)|\, dx$. Since g is continuous and f is not, we know that $f \neq g$. If f were continuous on $[0, 1]$, then we could deduce that

$$d(f, g) = \int_0^1 |g(x) - f(x)|\, dx > 0.$$

Unfortunately, we cannot apply the metric d since f is not continuous and so is not in $C[0, 1]$. Both f restricted to $[0, \tfrac{1}{2})$ and f restricted to $[\tfrac{1}{2}, 1]$ are continuous, and so we consider the metrics on $C[0, \tfrac{1}{2})$ and $C[\tfrac{1}{2}, 1]$ given respectively by

$$d_-(g, h) = \int_0^{\frac{1}{2}} |h(x) - g(x)|\, dx \quad \text{and} \quad d_+(g, h) = \int_{\frac{1}{2}}^1 |h(x) - g(x)|\, dx.$$

The proof that d_- and d_+ are metrics is similar to the proof that d is a metric.

We know that there exists $x_0 \in [0, 1] - \{\tfrac{1}{2}\}$ such that $f(x_0) \neq g(x_0)$ and so one of $d_-(f, g)$ and $d_+(f, g)$ is positive. Since

$$\int_0^1 |g(x) - f(x)|\, dx = d_-(f, g) + d_+(f, g),$$

we have $\int_0^1 |g(x) - f(x)| > 0$, as required.

Thus $(C[0, 1], d)$ is not a complete metric space. ∎

This raises the question of whether we can find a complete metric space (X, d_X) that contains $(C[0, 1], d)$ as a subspace. In fact, we can. In Section 4, we discuss a method for constructing the *smallest* such space — the *completion* of $(C[0, 1], d)$.

2.3 Completeness is not a topological invariant

The following example illustrates that completeness is not a topological invariant.

Example 2.3

Let $X = (-1, 1)$ and $\mathbb{R}$ both have the Euclidean topology. The function $f\colon X \to \mathbb{R}$ given by $f(x) = \tan(\pi x/2)$ is a homeomorphism, so $(-1, 1)$ is homeomorphic to $\mathbb{R}$ when they both carry their Euclidean topologies.

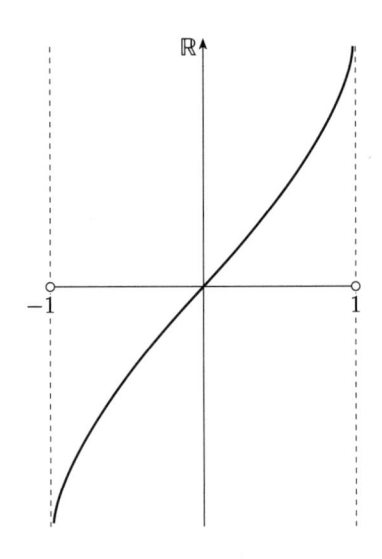

Figure 2.3 A homeomorphism between $(-1, 1)$ and $\mathbb{R}$.

Now consider the sequence $(1 - 1/n)$ in X. It is a Cauchy sequence since, if $N \in \mathbb{N}$ and $m > n > N$, then
$$\left(1 - \frac{1}{m}\right) - \left(1 - \frac{1}{n}\right) = \frac{1}{n} - \frac{1}{m} < \frac{2}{N} \to 0, \text{ as } m, n \to \infty.$$
This sequence converges to 1 in $\mathbb{R}$ but does not converge in X, since $1 \notin (-1, 1)$. Hence $(X, d^{(1)})$ is incomplete. So we have two homeomorphic metric spaces, one complete and the other not: thus completeness is not a topological invariant. ∎

We gain some insight into the above example by noting that if we apply the homeomorphism f to the terms of the sequence $(1 - 1/n)$ in $(-1, 1)$, we obtain the image sequence $(\tan(\pi/2 - \pi/2n))$ in $\mathbb{R}$. This is not a Cauchy sequence in $\mathbb{R}$: for if it were, it would converge (since $\mathbb{R}$ is complete) and its limit would be $\tan \pi/2$, which is undefined.

Although completeness is not a topological invariant, it is a metric invariant:

> *if two metric spaces are metrically the same, then they are either both complete or both incomplete.*

In order to make sense of this statement, we need to understand what it means for two metric spaces to be 'metrically the same'. The key property is the notion of distance: two metric spaces are the same if we can map the points of one space to the points of the other space so that all distance relationships are preserved. The following definition makes this precise.

Definition

Let (X, d_X) and (Y, d_Y) be metric spaces. A function $f: X \to Y$ is an **isometry** if:

(a) f is one–one and onto;
(b) for each $x_1, x_2 \in X$, $d_X(x_1, x_2) = d_Y(f(x_1), f(x_2))$.

The metric spaces (X, d_X) and (Y, d_Y) are **isometric** if there is an isometry between X and Y.

An isometry is sometimes called a *distance-preserving function*.

Remarks

(i) If $f: X \to Y$ is an isometry, then it is invertible (since it is one–one and onto) and its inverse f^{-1} is also an isometry.

(ii) The notion of 'is isometric to' is an equivalence relation on metric spaces.

See the Exercises for this unit.

(iii) Since an isometry preserves distances, a sequence in X converges if and only if its image under the isometry converges in Y. Equivalently, a sequence in X is a d_X-Cauchy sequence if and only if its image under the isometry is a d_Y-Cauchy sequence in Y.

See Problem 2.6.

(iv) Since closed sets and compact sets can be characterized in terms of sequences, an isometry preserves closed and compact sets, and thus also preserves open sets. This implies that connectedness is also preserved under an isometry.

Problem 2.5

Show that:

(a) $([0,1], d^{(1)})$ is isometric to $([1,2], d^{(1)})$;

(b) $([0,1], d^{(1)})$ is not isometric to $([0,2], d^{(1)})$.

Problem 2.6

Let (X, d_X) and (Y, d_Y) be metric spaces, and let $f\colon X \to Y$ be an isometry. Show that if (a_n) is a sequence in X that converges to a, then $(f(a_n))$ converges to $f(a)$ in Y.

Problem 2.5(b) demonstrates that homeomorphic metric spaces are not necessarily isometric. Thus the following theorem does not contradict Example 2.3.

Theorem 2.6

Let (X, d_X) and (Y, d_Y) be isometric metric spaces. If (X, d_X) is complete, then (Y, d_Y) is also complete.

Proof Suppose that (X, d_X) and (Y, d_Y) are isometric metric spaces and let $f\colon X \to Y$ be an isometry. Let (X, d_X) be complete and let (a_n) be a Cauchy sequence in Y. Then $(f^{-1}(a_n))$ is a Cauchy sequence in X since, for any $n, m \in \mathbb{N}$,

$$d_X(f^{-1}(a_n), f^{-1}(a_m)) = d_Y(a_n, a_m).$$

Since f is an isometry, it is invertible.

Since (X, d_X) is complete, there is an $a \in X$ such that $(f^{-1}(a_n))$ converges to a. But

$$d_Y(a_n, f(a)) = d_X(f^{-1}(a_n), a) \to 0, \text{ as } n \to \infty,$$

so (a_n) converges to $f(a)$ in Y. Hence (Y, d_Y) is complete. ∎

Isometries play a big role in Section 4, where we discuss how to form the *completion* of a metric space — the smallest complete metric space that contains the given space.

3 The Contraction Mapping Theorem

> After working through this section, you should be able to:
> ▶ appreciate the role that fixed-point theorems play in many areas of mathematics;
> ▶ determine whether a given function is a contraction mapping;
> ▶ use the Contraction Mapping Theorem to solve simple fixed-point problems.

In this section we show how complete metric spaces play an important role in the theory of fixed-points of mappings. Finding the fixed-points of a particular function (those points that are mapped to themselves) is frequently an important part of solving a mathematical problem.

We give a formal definition of fixed-points below.

3.1 Fixed-points and the Contraction Mapping Theorem

Many problems in mathematics can be written in the form:

solve the equation $f(x) = x$.

Such points x are known as *fixed-points*.

Definition

Let X be a set and let $T: X \to X$ be a function.

A **fixed-point** of T is a point $x \in X$ such that $T(x) = x$.

We use T to denote the function, not f, because the set X may have functions as its elements.

Fixed-points are surprisingly important and many theorems provide sufficient conditions for a function T to have a fixed-point. In this section we prove the Contraction Mapping Theorem, which ensures that, under appropriate conditions, a function has a *unique* fixed-point. This theorem also provides a good method for finding an approximate value for the fixed-point.

We shall see some examples shortly.

Problem 3.1

Let $f: \mathbb{R} \to \mathbb{R}$ be a function and define $g: \mathbb{R} \to \mathbb{R}$ by $g(x) = x - f(x)$. Show that x is a fixed-point of g if and only if x is a zero of f — that is, $f(x) = 0$.

Thus the problem of finding the zeros of a real-valued function can be converted into a problem concerning the fixed-points of a (different) real-valued function. For example, finding the zeros of

$\cos x - \exp x$ for $x \in \mathbb{R}$,

See Figure 3.1.

is the same as finding the fixed-points of the function $g: \mathbb{R} \to \mathbb{R}$ given by

$g(x) = x - \cos x + \exp x$.

See Figure 3.2.

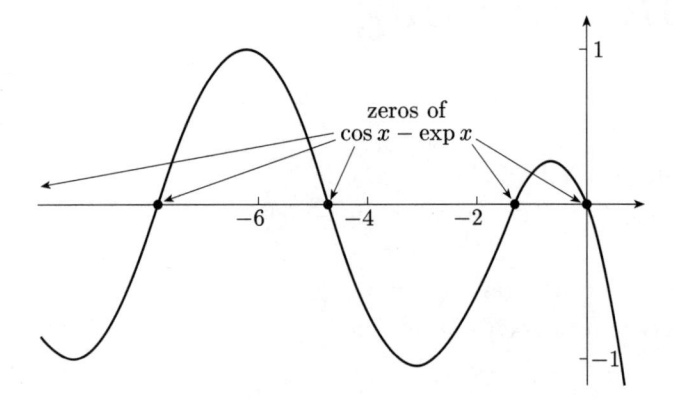

Figure 3.1 Zeros of $\cos x - \exp x$.

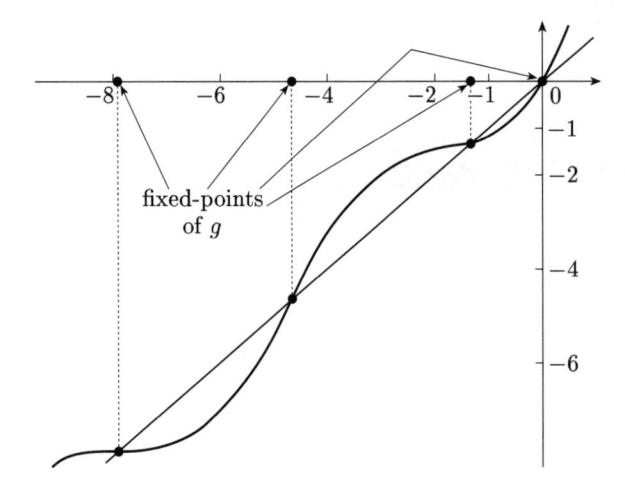

Figure 3.2 Fixed-points of g.

Our second example of a fixed-point problem involves a mapping defined on $C[0, 1]$.

Example 3.1

Consider the differential equation

$$\frac{df}{ds} = sf(s), \quad f(0) = 1 \quad (0 \leq s \leq 1). \tag{3.1}$$

Integrating both sides of this equation with respect to s gives

$$\int_0^t \frac{df}{ds}\, ds = \int_0^t sf(s)\, ds$$

and so

$$f(t) - f(0) = \int_0^t sf(s)\, ds \quad \text{for } 0 \leq t \leq 1.$$

Using the boundary condition $f(0) = 1$, we obtain an *integral equation* that any solution f of the differential equation must satisfy:

$$f(t) = 1 + \int_0^t sf(s)\, ds \quad \text{for } 0 \leq t \leq 1.$$

Let us define a mapping T by

$$T(f)(t) = 1 + \int_0^t sf(s)\, ds \quad \text{for } f \in C[0, 1]; \tag{3.2}$$

then any solution of the differential equation (3.1) is also a fixed-point of this mapping T. We now show that T is a mapping from $C[0, 1]$ to itself.

First, note that $s \mapsto sf(s)$ is continuous, since it is the product of continuous functions and so belongs to $C[0,1]$. Also, the mapping F defined by

$$F(g)(t) = \int_0^t g(s)\,ds \quad \text{for } 0 \le t \le 1,$$

is a mapping from $C[0,1]$ to itself. So the function G defined by

$$G(f)(t) = \int_0^t sf(s)\,ds \quad \text{for } 0 \le t \le 1,$$

being the composition of two continuous functions, is in $C[0,1]$. The constant function 1 is also in $C[0,1]$, and so T is a mapping from $C[0,1]$ to itself. ∎

This is shown in Exercise A2.9.

Problem 3.2

Consider the differential equation

$$\frac{df}{ds} = \tfrac{1}{2}\cos f(s), \quad f(0) = 0 \quad (0 \le s \le 1).$$

By integrating both sides of this equation with respect to s, find an integral equation that every solution of this differential equation must satisfy.

Hence define a mapping $T: C[0,1] \to C[0,1]$ such that any solution of the differential equation is also a fixed-point of this mapping.

The theorem we now aim to prove is the *Contraction Mapping Theorem*. This theorem guarantees the existence of a fixed-point when the mapping from a complete metric space to itself is of a special type, known as a *contraction mapping*.

This is also known as the *Banach Fixed-Point Theorem* after the Polish mathematician Stefan Banach (1892–1945). Banach was a prolific analyst and has many concepts and theorems named after him.

Definition

Let (X, d) be a metric space. A function $T: X \to X$ is a **contraction mapping** if there is a real number $0 \le \lambda < 1$ such that

$$d(T(x), T(y)) \le \lambda d(x, y) \quad \text{for all } x, y \in X.$$

Any such real number λ is a **contraction ratio** for T.

Remarks

(i) A contraction mapping is a Lipschitz function from a set X to itself with Lipschitz constant *strictly less than* 1. For example, $T: \mathbb{R} \to \mathbb{R}$ given by $T(x) = \tfrac{1}{2}x$ is a Lipschitz mapping with Lipschitz constant $\tfrac{1}{2}$ for the Euclidean metric, and so is a contraction mapping.

(ii) A contraction mapping has many different possible contraction ratios: for if $\lambda \in (0,1)$ is a contraction ratio, then so is any $\mu \in (\lambda, 1)$.

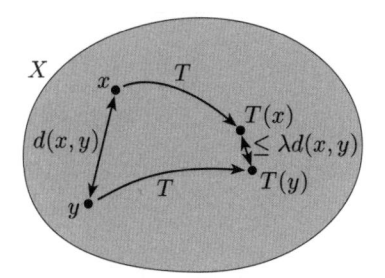

Figure 3.3

Worked problem 3.1

Show that for the metric $d_{\max}$ on $C[0,1]$, the mapping $T: C[0,1] \to C[0,1]$ defined by

$$T(f)(t) = 1 + \int_0^t sf(s)\,ds \quad \text{for } 0 \le t \le 1,$$

is a contraction mapping with contraction ratio $\tfrac{1}{2}$.

You met this mapping in Example 3.1.

Solution

Given $f, g \in C[0,1]$ and $t \in [0,1]$, we have

$$\begin{aligned}|T(f)(t) - T(g)(t)| &= \left|\int_0^t s(f(s) - g(s))\,ds\right| \\ &\leq \int_0^t |s|\,|f(s) - g(s)|\,ds \\ &\leq d_{\max}(f,g) \int_0^t s\,ds \\ &= d_{\max}(f,g)\tfrac{1}{2}t^2 \leq \tfrac{1}{2}d_{\max}(f,g).\end{aligned}$$

Since t is arbitrary, we deduce that

$$d_{\max}(T(f), T(g)) \leq \tfrac{1}{2}d_{\max}(f,g),$$

and so T is a contraction mapping with contraction ratio $\tfrac{1}{2}$. ∎

Problem 3.3

Let (X, d) be a metric space and let $T\colon X \to X$ be a contraction mapping with contraction ratio 0. Show that T is a *constant* function.

Problem 3.4

Let $T\colon C[0,1] \to C[0,1]$ be defined by

$$T(f)(t) = \int_0^t \tfrac{1}{2}\cos f(s)\,ds \quad \text{for } 0 \leq t \leq 1.$$

You met this mapping in Problem 3.2.

Show that T is a contraction mapping with contraction ratio $\tfrac{1}{2}$.

Hint Use the fact that $|\cos a - \cos b| \leq |a - b|$ for any real numbers a and b.

We now show that contraction mappings can have at most one fixed-point. We use this result in the proof of the Contraction Mapping Theorem.

> **Lemma 3.1**
>
> Let (X, d) be a metric space. Suppose that $T\colon X \to X$ is a contraction mapping. Then T has at most one fixed-point.

Proof Suppose that $x, y \in X$ are fixed-points of T. Then

$$T(x) = x \quad \text{and} \quad T(y) = y.$$

Since T is a contraction mapping,

$$d(x, y) = d(T(x), T(y)) \leq \lambda d(x, y),$$

where $0 \leq \lambda < 1$. Rearranging, we obtain

$$(1 - \lambda)d(x, y) \leq 0,$$

which implies that $d(x, y) \leq 0$, since $0 \leq \lambda < 1$. But $d(x, y) \geq 0$, and so $d(x, y) = 0$. Hence since d is a metric, $x = y$. Thus T has at most one fixed-point. ∎

Remark

The conclusion of this lemma can fail if $T: X \to X$ is a Lipschitz mapping with Lipschitz constant 1 — that is,

$$d(T(x), T(y)) \leq d(x, y) \quad \text{for } x, y \in X.$$

For example, consider $T: \mathbb{R} \to \mathbb{R}$ (with the usual metric), where $T(x) = x$. For this mapping, *every* real number is a fixed-point. This is why we require a contraction mapping to have a Lipschitz constant *strictly* less than one — it guarantees that the mapping can have at most one fixed-point.

Problem 3.5

Let $(0, 1]$ have the Euclidean metric, and define $T: (0, 1] \to (0, 1]$ by $T(x) = \frac{1}{2}x$. Show that:

(a) T is a contraction mapping with contraction ratio $\frac{1}{2}$;

(b) T has no fixed-points.

Problem 3.5 shows that a contraction mapping need not have any fixed-points: the difficulty is that the point that 'should be' the fixed-point of the mapping is not in the space. This difficulty cannot arise for a *complete* metric space.

Theorem 3.2 Contraction Mapping Theorem

Let (X, d) be a complete metric space, and let $T: X \to X$ be a contraction mapping. Then there is a unique element $x_T \in X$ such that

$$T(x_T) = x_T.$$

Moreover, for each $x \in X$, the sequence of iterates

$$x, T(x), T(T(x)), T(T(T(x))), \ldots$$

converges to this unique fixed-point x_T — that is, if

$$T^n(x) = (T \circ T \circ \cdots \circ T)(x) \quad (T \text{ composed with itself } n \text{ times}),$$

then

$$d(T^n(x), x_T) \to 0 \quad \text{as } n \to \infty,$$

for any $x \in X$.

Proof Our proof involves three steps.

(a) We show that, for each $x \in X$, the sequence $(x, T(x), T^2(x), \ldots)$ is a Cauchy sequence. Since (X, d) is complete, this sequence converges to some element $x_T \in X$.

(b) We verify that $T(x_T) = x_T$ — that is, x_T is a fixed-point of T.

(c) We deduce from Lemma 3.1 that the fixed-point is unique.

Since T is a contraction mapping, there is a number λ (with $0 \leq \lambda < 1$) such that

$$d(T(x), T(y)) \leq \lambda d(x, y) \quad \text{for all } x, y \in X.$$

Now fix $x \in X$ and consider the sequence of iterates $x, T(x), T^2(x), \ldots$.

(a) *Show that $(T^n(x))$ is a Cauchy sequence*

We need to show that, for each $\varepsilon > 0$, there is an $N \in \mathbb{N}$ such that
$$d(T^n(x), T^m(x)) < \varepsilon \quad \text{for all } n, m > N.$$

We first estimate $d(T^k(x), T^{k+1}(x))$, the distance between two successive terms of the sequence:

$$\begin{aligned}
d(T^k(x), T^{k+1}(x)) &= d(T(T^{k-1}(x)), T(T^k(x))) \\
&\leq \lambda d(T^{k-1}(x), T^k(x)), \quad \text{since } T \text{ is a contraction mapping,} \\
&= \lambda d(T(T^{k-2}(x)), T(T^{k-1}(x))) \\
&\leq \lambda^2 d(T^{k-2}(x), T^{k-1}(x)), \quad \text{since } T \text{ is a contraction mapping,} \\
&\vdots \quad \text{(repeating this } k \text{ times)} \\
&\leq \lambda^k d(x, T(x)). \quad (3.3)
\end{aligned}$$

Let n, m be integers with $m > n$. By (3.3) and repeated use of the Triangle Inequality, we find:

$$\begin{aligned}
&d(T^n(x), T^m(x)) \\
&\leq d(T^n(x), T^{n+1}(x)) + d(T^{n+1}(x), T^{n+2}(x)) + \cdots + d(T^{m-1}(x), T^m(x)) \\
&\leq \lambda^n d(x, T(x)) + \lambda^{n+1} d(x, T(x)) + \cdots + \lambda^{m-1} d(x, T(x)) \\
&= (\lambda^n + \lambda^{n+1} + \lambda^{n+2} + \cdots + \lambda^{m-1}) d(x, T(x)) \\
&= \frac{\lambda^n - \lambda^m}{1 - \lambda} d(x, T(x)) \\
&\leq \frac{\lambda^n}{1 - \lambda} d(x, T(x)). \quad (3.4)
\end{aligned}$$

Let $\varepsilon > 0$. If we choose an $N \in \mathbb{N}$ so that
$$\frac{\lambda^N}{1-\lambda} d(x, T(x)) \leq \varepsilon,$$
then, for all $m > n > N$,
$$d(T^n(x), T^m(x)) \leq \frac{\lambda^n}{1-\lambda} d(x, T(x)) < \frac{\lambda^N}{1-\lambda} d(x, T(x)) \leq \varepsilon,$$
since $\lambda^n < \lambda^N$ for $\lambda < 1$. Thus the sequence is a Cauchy sequence.

Since (X, d) is a complete metric space, the sequence $(T^n(x))$ converges to some point $x_T \in X$.

(b) *Verify that x_T is a fixed-point of T*

By the Triangle Inequality, we have, for each $n \geq 1$:
$$\begin{aligned}
d(x_T, T(x_T)) &\leq d(x_T, T^n(x)) + d(T^n(x), T(x_T)) \\
&\leq d(x_T, T^n(x)) + \lambda d(x_T, T^{n-1}(x)).
\end{aligned}$$

Since n is arbitrary, and $d(x_T, T^n(x)) \to 0$ as $n \to \infty$, we can take the limit as $n \to \infty$ on the right-hand side. We conclude that
$$d(x_T, T(x_T)) \leq 0 + \lambda \cdot 0 = 0,$$
that is, $T(x_T) = x_T$.

(c) *Show that the fixed-point is unique*

This follows immediately from Lemma 3.1. ∎

Worked problem 3.2

Use the Contraction Mapping Theorem to show that the equation $x = \frac{1}{2} \cos x$ has a unique real solution.

In this Worked problem, we could use the Intermediate Value Theorem to show that a root exists, but that approach does not show that there is just one root.

Solution

This problem does not specify which metric space we should work in. We choose a complete metric space and an appropriate contraction mapping T. In this case, there is a natural choice of mapping and space.

> This is common in situations involving the Contraction Mapping Theorem.

Let $T: \mathbb{R} \to \mathbb{R}$ be given by $T(x) = \frac{1}{2}\cos x$, and consider the complete metric space $(\mathbb{R}, d^{(1)})$. Note that $T(x) = x$ if and only if $x = \frac{1}{2}\cos x$, and so the roots of this equation correspond exactly to the fixed-points of T. It remains only to check that T is a contraction mapping. For $x, y \in \mathbb{R}$, the Mean Value Theorem implies that there exists $c \in (x, y)$ such that

$$|T(x) - T(y)| = |\tfrac{1}{2}(\cos x - \cos y)| = |T'(c)|\,|x - y|.$$

Now

$$|T'(c)| = |\tfrac{1}{2}\sin c| \leq \tfrac{1}{2}, \text{ for } c \in \mathbb{R},$$

and so T is a contraction mapping with contraction ratio $\frac{1}{2}$. By the Contraction Mapping Theorem, T has a unique fixed-point and so the original equation has a unique real solution. ∎

Problem 3.6

Define $T: [1, \infty) \to [1, \infty)$ by

$$T(x) = x + \frac{1}{x}.$$

Show that for all $x, y \in [1, \infty)$ with $x \neq y$,

$$|T(x) - T(y)| < |x - y|.$$

Does this mapping have a fixed-point? If not, does this contradict the Contraction Mapping Theorem?

The proof of the Contraction Mapping Theorem can also be used to obtain an estimate of how far $T^n(x)$ is from x_T (the unique fixed-point), in terms of $d(x, T(x))$. This is useful when we use a computer to find an approximate value for the fixed-point.

Corollary 3.3

Let (X, d) be a complete metric space and let $T: X \to X$ be a contraction mapping with contraction ratio $\lambda \in [0, 1)$. Then, for each $x \in X$ and each $n \in \mathbb{N}$,

$$d(x_T, T^n(x)) \leq \frac{\lambda^n}{1 - \lambda} d(x, T(x)),$$

where x_T is the fixed-point of T.

Proof Let $x \in X$ and take $m, n \in \mathbb{N}$ with $m > n$. By the Triangle Inequality,

$$d(x_T, T^n(x)) \leq d(x_T, T^m(x)) + d(T^m(x), T^n(x))$$

$$\leq d(x_T, T^m(x)) + \frac{\lambda^n}{1 - \lambda} d(x, T(x)), \quad \text{by (3.4)}.$$

By the Contraction Mapping Theorem, $d(x_T, T^m(x)) \to 0$ as $m \to \infty$, and so

$$d(x_T, T^n(x)) \leq \frac{\lambda^n}{1 - \lambda} d(x, T(x)). \qquad \blacksquare$$

Consider the mapping $T: \mathbb{R} \to \mathbb{R}$ given by $T(x) = \frac{1}{2}\cos x$. In the solution to Worked problem 3.2 we showed that T is a contraction mapping with contraction ratio $\frac{1}{2}$ and so, by the Contraction Mapping Theorem, T has a unique fixed-point x_T. We use Corollary 3.3 to estimate the value of x_T.

Taking $n = 1$ and $x = 0$, we have

$$|x_T - T(0)| \leq \frac{\frac{1}{2}}{1 - \frac{1}{2}}|0 - T(0)| = \frac{1}{2}.$$

Since $T(0) = \frac{1}{2}$, the root lies somewhere in $[0, 1]$. We can use Corollary 3.3 to obtain finer approximations to x_T by looking at $T^n(0)$ for larger values of n. For example, taking $n = 2$, we obtain

$$|x_T - T^2(0)| \leq \frac{(\frac{1}{2})^2}{1 - \frac{1}{2}}|0 - T(0)| = \frac{1}{4}.$$

Since $T^2(0) = \frac{1}{2}\cos\frac{1}{2}$, this implies that x_T lies in the interval

$$[\tfrac{1}{2}\cos\tfrac{1}{2} - \tfrac{1}{4}, \tfrac{1}{2}\cos\tfrac{1}{2} + \tfrac{1}{4}] \subseteq [0.188, 0.689].$$

Problem 3.7

Consider the mapping $T: \mathbb{R} \to \mathbb{R}$ given by $T(x) = \frac{1}{2}\cos x$. Find an interval of length at most 0.1 containing the unique fixed-point of T.

3.2 Applications of the Contraction Mapping Theorem

We now give some applications of the Contraction Mapping Theorem.

In *Unit C5*, we use the Contraction Mapping Theorem to define fractals.

Locating zeros of real functions

If a continuous real-valued function on an interval is a contraction mapping, then we can use the Contraction Mapping Theorem to deduce the existence of a unique fixed-point in that interval, and Corollary 3.3 to estimate how far we are from this fixed-point when we iterate an initial guess.

Unfortunately, many of the functions whose zeros we wish to find cannot be directly rearranged to give a fixed-point problem involving contractions. For example, suppose that we wish to find the zeros of

$$x^2 - e^{-x}.$$

Following Problem 3.1, it is natural to look for the fixed-points of the function $g: \mathbb{R} \to \mathbb{R}$ given by $g(x) = x - x^2 + e^{-x}$. But g is not a contraction mapping, and so we cannot apply the Contraction Mapping Theorem *directly*. However, by carefully defining the auxiliary function g, we *can* use the Contraction Mapping Theorem to find good approximations to the zeros of f.

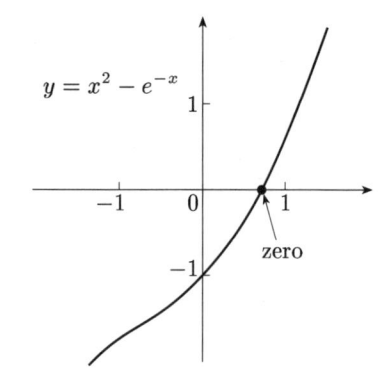

Figure 3.4 A sketch of the graph of f suggests that there is only one zero.

Suppose we wish to find a zero of $f: \mathbb{R} \to \mathbb{R}$ given by $f(x) = x^2 - e^{-x}$. The function f is continuous, $f(0) = -1$ and $f(1) = 1 - e^{-1} > 0$. By the Intermediate Value Theorem, there is a zero of f in the interval $[0, 1]$.

This method tells us that there is a zero in the interval $[0, 1]$ but it does not tell us whether there are any zeros outside this interval. We need to use other methods to decide this.

The idea is to notice that the zeros of f correspond to the fixed-points of the function g_r given by $g_r(x) = x - rf(x)$, for *any* non-zero real number r. Thus if we can find a non-zero real number r and a closed interval I such that $g_r: I \to I$ and g_r is a contraction mapping on this interval, then we can deduce from the Contraction Mapping Theorem that f has a unique zero in this interval.

With this objective in mind, we estimate the derivative of g_r on the interval $[0, 1]$, where f has a zero. We have $f'(x) = 2x + e^{-x}$, and so

$$1 \leq f'(x) \leq 3 \quad \text{for all } x \in [0, 1].$$

This tells us that f is strictly increasing on the interval $[0, 1]$.

The function g_r has derivative

$$g_r'(x) = 1 - rf'(x) \quad \text{for all } x \in [0, 1].$$

Hence for any $r > 0$,

$$1 - 3r \leq g_r'(x) \leq 1 - r \quad \text{for all } x \in [0, 1].$$

Moreover, if $r \leq \frac{1}{3}$, then $1 - 3r \geq 0$ and so g_r' is non-decreasing on $[0, 1]$. Hence for $0 \leq x \leq 1$ and $0 < r \leq \frac{1}{3}$,

$$g_r(x) \geq g_r(0) = 0 - rf(0) > 0$$

and

$$g_r(x) \leq g_r(1) = 1 - rf(1) < 1,$$

and so $g_r: [0, 1] \to [0, 1]$. Moreover, for $0 < r \leq \frac{1}{3}$, it follows from the Mean Value Theorem that, for all $x, y \in [0, 1]$,

$$|g_r(x) - g_r(y)| \leq \max_{0 \leq c \leq 1} |g_r'(c)| \, |x - y| \leq (1 - r)|x - y|.$$

Thus for $0 < r \leq \frac{1}{3}$, g_r is a contraction mapping on $[0, 1]$ with contraction ratio $1 - r$. By the Contraction Mapping Theorem, g_r has a unique fixed-point in the interval $[0, 1]$ and thus f has a unique zero in $[0, 1]$. If $r = \frac{1}{3}$ and x_r is the fixed-point of g_r (and the zero of f) then, by Corollary 3.3,

$$|x_r - g_r^n(x_0)| \leq \frac{(1-r)^n}{1 - (1-r)} |x_0 - g_r(x_0)| = 3(\tfrac{2}{3})^n |x_0 - g_r(x_0)| \leq 3(\tfrac{2}{3})^n,$$

whenever $x_0 \in [0, 1]$ is an initial guess for the value of the fixed-point.

This method finds the zeros of any function f that is strictly monotonic in a neighbourhood of the zero being sought.

Figure 3.5: f is strictly increasing on $[0, 1]$, $1 \leq f'(x) \leq 3$

Theorem 3.4

Let $a < b \in \mathbb{R}$, and suppose that $f: [a, b] \to \mathbb{R}$ is continuous and differentiable on $[a, b]$. If
(a) $f(a) < 0 < f(b)$,
(b) there are $m, M \in \mathbb{R}$ with $0 < m \leq M$ such that

$$m \leq f'(x) \leq M \quad \text{for all } x \in [a, b],$$

then, for $0 < r \leq \frac{1}{M}$, the function $g_r: [a, b] \to \mathbb{R}$ given by $g_r(x) = x - rf(x)$ maps $[a, b]$ into itself, and is a contraction mapping on $[a, b]$ with contraction ratio $1 - mr$.

It follows from the Contraction Mapping Theorem that a function f satisfying the conditions of Theorem 3.4 has a unique zero in $[a, b]$. We can then use Corollary 3.3 (applied to g_r) to find the approximate location of this zero.

Problem 3.8

Prove Theorem 3.4.

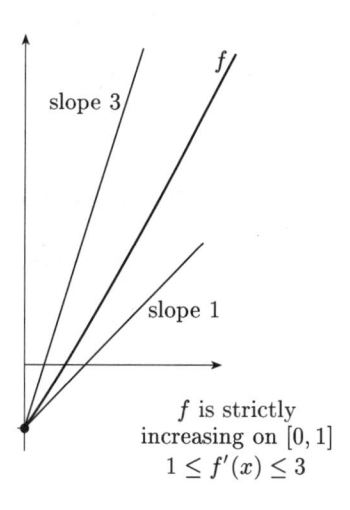

Solving differential equations

Recall from Example 3.1 that any solution of the differential equation

$$\frac{df}{ds} = sf(s), \quad f(0) = 1 \quad (0 \leq s \leq 1)$$

must also satisfy the integral equation

$$f(t) = 1 + \int_0^t sf(s)\,ds \quad \text{for } 0 \leq t \leq 1.$$

We wish to find the fixed-points of the mapping $T: C[0,1] \to C[0,1]$, defined in equation (3.2) by

$$T(f)(t) = 1 + \int_0^t sf(s)\,ds \quad \text{for } 0 \leq t \leq 1.$$

In Worked problem 3.1, we saw that, for $d_{\max}$, T is a contraction mapping with contraction ratio $\frac{1}{2}$. In Section 2, we proved that $(C[0,1], d_{\max})$ is a complete metric space. So, by the Contraction Mapping Theorem, there is a unique function $f \in C[0,1]$ such that $f = T(f)$ — that is,

$$f(t) = 1 + \int_0^t sf(s)\,ds. \tag{3.5}$$

We now find approximations to the fixed-point of T, by using Corollary 3.3.

We start by calculating $T(f_0)$ for a simple function $f_0 \in C[0,1]$ — say $f_0(t) = 1$ for all $t \in [0,1]$, so that

$$T(f_0)(t) = 1 + \int_0^t s\,ds = 1 + \tfrac{1}{2}t^2.$$

Hence

$$|T(f_0)(t) - f_0(t)| = \tfrac{1}{2}t^2,$$

and so

$$d_{\max}(T(f_0), f_0) = \max\{\tfrac{1}{2}t^2 : t \in [0,1]\} = \tfrac{1}{2}.$$

Thus if $f \in C[0,1]$ is the fixed-point of T, then by Corollary 3.3,

$$d_{\max}(f, T^n(f_0)) \leq \frac{(\tfrac{1}{2})^n}{1 - \tfrac{1}{2}} d_{\max}(T(f_0), f_0) = (\tfrac{1}{2})^n.$$

We now calculate $T^2(f_0)$ for $t \in [0,1]$: we have

$$T^2(f_0)(t) = T(T(f_0))(t)$$
$$= 1 + \int_0^t s(1 + \tfrac{1}{2}s^2)\,ds$$
$$= 1 + \tfrac{1}{2}t^2 + \tfrac{1}{8}t^4 = 1 + (\tfrac{1}{2}t^2) + \tfrac{1}{2}(\tfrac{1}{2}t^2)^2.$$

Similarly, we find $T^3(f_0)$ for $t \in [0,1]$: we have

$$T^3(f_0)(t) = 1 + \int_0^t s\left(1 + (\tfrac{1}{2}s^2) + \tfrac{1}{2}(\tfrac{1}{2}s^2)^2\right) ds$$
$$= 1 + (\tfrac{1}{2}t^2) + \tfrac{1}{2!}(\tfrac{1}{2}t^2)^2 + \tfrac{1}{3!}(\tfrac{1}{2}t^2)^3.$$

In fact, using the identity

$$\int_0^t s\left(\tfrac{1}{2}s^2\right)^k ds = \frac{t^{2(k+1)}}{2^k(2k+2)} = \frac{1}{k+1}\left(\tfrac{1}{2}t^2\right)^{k+1},$$

we can use mathematical induction to conclude that

$$T^n(f_0)(t) = \sum_{k=0}^{n} \frac{1}{k!} \left(\tfrac{1}{2}t^2\right)^k.$$

Since convergence for the $d_{\max}$ metric implies pointwise convergence, the fixed-point f of T is given by

Unit C3, Theorem 3.3 and Lemma 3.1.

$$f(t) = \lim_{n \to \infty} T^n(f_0)(t) = \sum_{k=0}^{\infty} \frac{1}{k!} \left(\tfrac{1}{2}t^2\right)^k = \exp(\tfrac{1}{2}t^2).$$

Thus $f(t) = \exp(\tfrac{1}{2}t^2)$ is the fixed-point of the mapping T.

Problem 3.9

Let $f(t) = \exp(\tfrac{1}{2}t^2)$. Verify that f satisfies (3.5).

We still need to check that the function $f(t) = \exp(\tfrac{1}{2}t^2)$ also satisfies the original differential equation

$$\frac{df}{ds} = sf(s), \quad f(0) = 1 \quad (0 \leq s \leq 1).$$

The following result guarantees that a solution to the integral equation arising from a given differential equation is also a solution of the original differential equation.

Theorem 3.5

Let $F \colon \mathbb{R}^2 \to \mathbb{R}$ be a $(d^{(2)}, d^{(1)})$-continuous function and let $a \in \mathbb{R}$. Then $f \in C[0,1]$ is a solution of the differential equation

$$\frac{df}{ds} = F(s, f(s)), \quad f(0) = a \quad (0 \leq s \leq 1),$$

if and only if

$$f(t) = a + \int_0^t F(s, f(s))\,ds \quad \text{for } 0 \leq t \leq 1.$$

We omit the proof.

Problem 3.10

Use the solutions to Problems 3.2 and 3.4 to show that the differential equation

$$\frac{df}{ds} = \tfrac{1}{2}\cos f(s), \quad f(0) = 0 \quad (0 \leq s \leq 1),$$

has a unique solution in $C[0,1]$.

By taking $f_0(s) = 0$, for $0 \leq s \leq 1$, as your initial approximation to this solution f, find an estimate $g \in C[0,1]$ of f for which

$$d_{\max}(f, g) \leq \tfrac{1}{4}.$$

Hint You will need to use Theorem 3.5 to show that you have a unique solution to the differential equation.

4 Completion

After working through this section, you should be able to:
▶ explain what is meant by the *completion* of a metric space;
▶ appreciate how to construct the completion of a given space.

In Section 3 we saw that in complete metric spaces, contraction mappings have a unique fixed-point, and that this is useful for solving a range of mathematical problems. In this section we show you how to find the 'smallest' complete metric space that contains a given metric space.

4.1 Defining the completion

Suppose that we are given a metric space (X, d) that is incomplete. Is it always possible to find a complete metric space that contains (X, d) as a subspace? This is an important question because, if it is not possible, the utility of the Contraction Mapping Theorem is restricted.

In Section 1, we saw that $((0, 1], d^{(1)})$ is not a complete metric space, but that $(\mathbb{R}, d^{(1)})$ is a complete metric space containing $((0, 1], d^{(1)})$. But $([0, 1], d^{(1)})$ is also a complete metric space that contains $((0, 1], d^{(1)})$ (since $[0, 1]$ is a closed subset of $\mathbb{R}$). Thus we have several possible choices for a complete metric space containing $((0, 1], d^{(1)})$, but it appears that $([0, 1], d^{(1)})$ is the 'smallest'.

Given a metric space containing (X, d), we aim to identify the 'smallest' complete metric space that contains (X, d) as a subspace. We begin by looking for a complete metric space (X^*, d^*) for which (X, d) is a *dense* subspace — that is, each d^*-open ball in X^* intersects X.

In the case of $((0, 1], d^{(1)})$, it is straightforward to identify such a metric space (X^*, d^*): it is $([0, 1], d^{(1)})$.

Problem 4.1

Find a complete metric space (X^*, d^*) for which $(\mathbb{Q}, d^{(1)})$ is a dense subspace.

If we can find a complete metric space (Y, d_Y) that contains (X, d) as a subspace, then we take $X^* = \mathrm{Cl}(X)$ and let d^* be the restriction of d_Y to X^* — Proposition 2.1 tells us that this is also a complete metric space. We then have a complete metric space that contains (X, d) as a dense set. Moreover, it is the *smallest* such space, for if $y \in \mathrm{Cl}(X) - X$ is omitted from X^*, then there is a sequence in X that converges to y and so does not converge in $X^* - \{y\}$.

For each of the metric spaces $((0, 1], d^{(1)})$ and $(\mathbb{Q}, d^{(1)})$, it is easy to find a complete metric space that contains it as a dense subspace, but in general it can be much harder to do this. For example, $(C[0, 1], d)$ is not a complete metric space when d is the integration metric given by

Section 2, Example 2.2.

$$d(f, g) = \int_0^1 |g(x) - f(x)|\, dx.$$

It seems difficult to identify a complete metric space that contains $(C[0,1], d)$. We seek a method that allows us to *construct* a complete metric space containing any given metric space as a dense subspace.

In Section 2, we introduced the notion of *isometry* and observed that isometric metric spaces are metrically the same: any properties that depend only on the metric are common to both spaces. In particular, we saw in Theorem 2.6 that completeness is an *isometric invariant*: if (X, d_X) and (Y, d_Y) are isometric, then one is complete if and only if the other is complete. Hence it is enough for us to construct a complete metric space (X^*, d^*) that contains a dense subspace that is isometric to (X, d). Such a metric space (X^*, d^*) is called a *completion* of (X, d).

> **Definition**
>
> Let (X, d) be a metric space. A **completion** of X is a complete metric space (X^*, d^*) that contains a dense subspace isometric to (X, d).

Remarks

(i) If (X^*, d^*) and (Y^*, e^*) are both completions of (X, d), then they are isometric — thus, up to isometry, completions are unique. Consequently, we usually talk about *the* completion of a space.   *We shall not prove this.*

(ii) Note that $([0,1], d^{(1)})$ is the completion of $((0,1], d^{(1)})$, and $(\mathbb{R}, d^{(1)})$ is the completion of $(\mathbb{Q}, d^{(1)})$.

Our task now is to find a procedure that allows us to construct the completion of an arbitrary metric space.

4.2 Constructing the completion of a space

Given a metric space (X, d), we aim to find a complete metric space (X^*, d^*) that contains a dense subspace isometric to (X, d). The key is to find a way of constructing a new metric space out of X that is 'bigger' than X and includes the limits of every Cauchy sequence in X.   *This section is not assessed.*

The idea is to use the Cauchy sequences in X as 'points' in the new space X^*, so that the Cauchy sequence $(a, a, a, \ldots)$ represents the point $a \in X$, and a Cauchy sequence that does not converge in X is the point in X^* that represents the limit of the sequence; for example, the sequence $(1, 1.4, 1.41, 1.414, 1.4142, \ldots)$, which is a Cauchy sequence in $(\mathbb{Q}, d^{(1)})$, represents $\sqrt{2}$.

Let X_C denote the collection of all the Cauchy sequences in X. So if $\mathbf{a} \in X_C$, then
$$\mathbf{a} = (a_1, a_2, a_3, a_4, \ldots), \quad \text{where } a_n \in X \text{ for each } n \in \mathbb{N},$$
and for each $\varepsilon > 0$, there is an $N \in \mathbb{N}$ such that
$$d(a_n, a_m) < \varepsilon \quad \text{for all } n, m > N.$$

Problem 4.2

When $(X, d) = (\mathbb{Q}, d^{(1)})$, which of the following are points in X_C?
(a) $(0, 0, 0, 0, \ldots)$ (b) $(1/n)_{n \in \mathbb{N}}$ (c) $(n)_{n \in \mathbb{N}}$

Unfortunately, the idea of defining the completion to be the set of all Cauchy sequences in X does not quite work. The problem is that two distinct Cauchy sequences may have the same limit, and so they should represent the same point.

We can perhaps better understand the problem by drawing an analogy with the problem of identifying $\{(p, q) : p, q \in \mathbb{Z}, q \neq 0\}$ with the set of rational numbers. We would like to say that if $p, q \in \mathbb{Z}$ with $q \neq 0$, then the pair (p, q) represents the rational number p/q. However this representation is not unique, since the pairs $(1, 2)$ and $(2, 4)$ both represent the rational number $\frac{1}{2}$. Instead, we define an equivalence relation on $\{(p, q) : p, q \in \mathbb{Z}, q \neq 0\}$ by defining $(p, q) \sim (p', q')$ if $p/q = p'/q'$, and we represent a rational number by the corresponding equivalence class.

We do the same here, and partition X_C into disjoint equivalence classes. A class consists of all the Cauchy sequences with the same 'limit' — or at least it would, if we had a limit to refer to! But the whole point of this construction is that we do not have *a priori* limits for all Cauchy sequences when (X, d) is incomplete.

So the next stage of the construction is to determine an equivalence relation on X_C that splits X_C into equivalence classes of Cauchy sequences that 'should' have the same limit. The following lemma helps us to do this.

Lemma 4.1

Let (X, d) be a metric space. If (a_n) and (b_n) are sequences in X that converge to the same limit, then $d(a_n, b_n) \to 0$ as $n \to \infty$.

Proof Let (a_n) and (b_n) be sequences in X that converge to the same limit $a \in X$. Let $\varepsilon > 0$; then

   there is an $N \in \mathbb{N}$ such that $d(a_n, a) < \varepsilon/2$, for all $n > N$,

and

   there is an $M \in \mathbb{N}$ such that $d(b_n, a) < \varepsilon/2$, for all $n > M$.

Hence whenever $n > \max\{N, M\}$, we have $d(a_n, a) < \varepsilon/2$ and $d(b_n, a) < \varepsilon/2$, and, by the Triangle Inequality,

$$d(a_n, b_n) \leq d(a_n, a) + d(a, b_n) < \varepsilon/2 + \varepsilon/2 = \varepsilon.$$

Thus $d(a_n, b_n) \to 0$ as $n \to \infty$.   ∎

We use this lemma to define a relation on X_C: if $\mathbf{a} = (a_n) \in X_C$ and $\mathbf{b} = (b_n) \in X_C$, then

   $\mathbf{a} \sim \mathbf{b}$ if and only if $d(a_n, b_n) \to 0$ as $n \to \infty$.

Problem 4.3

Show that $\sim$ is an equivalence relation on X_C.

We now define X^* to be the collection of equivalence classes of $(X_C, \sim)$.

We still need to define a metric d^* on X^* and to show that (X, d) is isometric to a dense subset of (X^*, d^*).

Let $\mathbf{a} = (a_1, a_2, a_3, \ldots)$ be a Cauchy sequence in X, and let $[\mathbf{a}]$ denote the equivalence class of $\mathbf{a}$. Thus

$$[\mathbf{a}] = \{\mathbf{a}' = (a_1', a_2', a_3', \ldots) \in X_C : d(a_n, a_n') \to 0 \text{ as } n \to \infty\}.$$

Given two elements $[\mathbf{a}], [\mathbf{b}] \in X^*$, we define our metric d^* as follows:

$$d^*([\mathbf{a}], [\mathbf{b}]) = \lim_{n \to \infty} d(a_n, b_n),$$

where $(a_n) \in [\mathbf{a}]$ and $(b_n) \in [\mathbf{b}]$.

First, we must verify that d^* is well defined — that is, that the limit exists, and if $\mathbf{a}' \in [\mathbf{a}]$ and $\mathbf{b}' \in [\mathbf{b}]$, then

$$\lim_{n \to \infty} d(a_n', b_n') = \lim_{n \to \infty} d(a_n, b_n).$$

For otherwise, d^* would depend on which representatives of the equivalence classes are used to calculate the limit, and so would not be well defined.

Lemma 4.2

Let $\mathbf{a} = (a_n) \in X_C$ and $\mathbf{b} = (b_n) \in X_C$. Then
(a) $\lim_{n \to \infty} d(a_n, b_n)$ exists;
(b) if $\mathbf{a}' = (a_n') \in [\mathbf{a}]$ and $\mathbf{b}' = (b_n') \in [\mathbf{b}]$, then

$$\lim_{n \to \infty} d(a_n', b_n') = \lim_{n \to \infty} d(a_n, b_n).$$

Proof

(a) To show that $\lim_{n \to \infty} d(a_n, b_n)$ exists, we prove that $(d(a_n, b_n))$ is a Cauchy sequence in $\mathbb{R}$. For $n, m \in \mathbb{N}$, we use the Triangle Inequality, followed by the Reverse Triangle Inequality, as follows:

$$0 \leq |d(a_n, b_n) - d(a_m, b_m)|$$
$$= |d(a_n, b_n) - d(a_m, b_n) + d(a_m, b_n) - d(a_m, b_m)|$$
$$\leq |d(a_n, b_n) - d(a_m, b_n)| + |d(a_m, b_n) - d(a_m, b_m)|$$
$$\leq d(a_n, a_m) + d(b_n, b_m).$$

But (a_n) and (b_n) are both Cauchy sequences, and so $(d(a_n, b_n))$ is also a Cauchy sequence. Since $(\mathbb{R}, d^{(1)})$ is complete, $(d(a_n, b_n))$ converges.

(b) Now suppose that $\mathbf{a}' \in [\mathbf{a}]$ and $\mathbf{b}' \in [\mathbf{b}]$.

To prove that d^* is defined independently of the chosen representatives of the equivalence classes, we show that

$$|d(a_n', b_n') - d(a_n, b_n)| \to 0 \text{ as } n \to \infty.$$

Since $\mathbf{a}' \in [\mathbf{a}]$ and $\mathbf{b}' \in [\mathbf{b}]$, we have $d(a_n', a_n) \to 0$ and $d(b_n', b_n) \to 0$. Thus by the Triangle Inequality and the Reverse Triangle Inequality

$$|d(a_n', b_n') - d(a_n, b_n)| = |d(a_n', b_n') - d(a_n', b_n) + d(a_n', b_n) - d(a_n, b_n)|$$
$$\leq |d(a_n', b_n') - d(a_n', b_n)| + |d(a_n', b_n) - d(a_n, b_n)|$$
$$\leq d(b_n', b_n) + d(a_n', a_n) \to 0 \quad \text{as } n \to \infty.$$

The result follows. ∎

Now we know that d^* is well defined, we can show that it defines a metric on X^*.

> **Theorem 4.3**
>
> (X^*, d^*) is a metric space.

Problem 4.4

Prove Theorem 4.3.

We next consider which elements of X^* can be identified with the points of X. If $x \in X$, then $(x, x, x, \ldots) \in X_C$ is a Cauchy sequence in X that converges to x. This gives us an embedding $i: X \to X^*$ where

$$i(x) = [(x, x, x, \ldots)].$$

Thus $i(x)$ is the equivalence class consisting of all Cauchy sequences (a_n) in X for which $d(a_n, x) \to 0$ — that is, $i(x)$ consists of all Cauchy sequences that converge to x.

Clearly, for all $x, y \in X$,

$$d^*(i(x), i(y)) = \lim_{n \to \infty} d(x, y) = d(x, y),$$

and so i is one–one; this implies that $i(X) = \{i(x) : x \in X\}$ is an isometric copy of X in X^*, and so (X, d) is isometric to $(i(X), d^*_{i(X)})$.

In fact, $i(X)$ is dense in X^*, as we now prove.

> **Lemma 4.4**
>
> The set $i(X)$ is a dense subset of (X^*, d^*).

Proof We need to show that, for each $a^* \in X^*$ and each $\varepsilon > 0$, there is an $x \in X$ for which $d^*(a^*, i(x)) < \varepsilon$.

So let $a^* \in X^*$ and let $\varepsilon > 0$.

Since $a^* \in X^*$, there is a Cauchy sequence $\mathbf{a} = (a_n)$ in X such that $a^* = [\mathbf{a}]$. Choose an $N \in \mathbb{N}$ such that $d(a_n, a_m) < \varepsilon/2$, for all $m, n > N$. Then, for $x = a_{N+1}$,

$$d^*(a^*, i(x)) = \lim_{n \to \infty} d(a_n, x) = \lim_{n \to \infty} d(a_n, a_{N+1}).$$

But, for $n > N$, $d(a_n, a_{N+1}) < \varepsilon/2$, and so

$$d^*(a^*, i(x)) < \varepsilon.$$

Thus $i(X)$ is dense in X^*. ∎

Finally, we prove that (X^*, d^*) is complete.

Theorem 4.5

The metric space (X^*, d^*) is complete.

Proof We show that Cauchy sequences in X^* converge to an element of X^*. Let $([\mathbf{a}]_m)$ be a Cauchy sequence in X^*. Each term in the sequence $([\mathbf{a}]_m)$ is itself a Cauchy sequence $(a_{m,n})_{n \in \mathbb{N}}$ in X.

Note that $([\mathbf{a}]_m)$ is a Cauchy sequence of Cauchy sequences.

We aim to identify a limit of the sequence $([\mathbf{a}]_m)$. To do this, we use Lemma 4.4 to find, for each $m \in \mathbb{N}$, a point $x_m \in X$ such that

$$d^*([\mathbf{a}]_m, i(x_m)) < \frac{1}{m}.$$

If we show that

(a) (x_m) is a Cauchy sequence in X,

(b) $d^*([\mathbf{a}]_m, [(x_j)]) \to 0$ as $m \to \infty$,

then we can deduce that $([\mathbf{a}]_m)$ is convergent (to $[(x_j)]$) in X^*, as required.

(a) *$(x_m)_{m \in \mathbb{N}}$ is a Cauchy sequence in X.*

Let $\varepsilon > 0$. For all $m, n \in \mathbb{N}$ and all $j \in \mathbb{N}$,

$$d(x_m, x_n) \leq d(x_m, a_{m,j}) + d(a_{m,j}, a_{n,j}) + d(a_{n,j}, x_n).$$

But

$$\lim_{j \to \infty} d(a_{m,j}, x_m) = d^*([\mathbf{a}]_m, i(x_m)) < \frac{1}{m}$$

and

$$\lim_{j \to \infty} d(a_{n,j}, x_n) = d^*([\mathbf{a}]_n, i(x_n)) < \frac{1}{n}.$$

Moreover, $([\mathbf{a}]_m)$ is a Cauchy sequence in X^*, and so there is an $N \in \mathbb{N}$ such that, for all $n, m > N$,

$$d^*([\mathbf{a}]_m, [\mathbf{a}]_n) = \lim_{j \to \infty} d(a_{m,j}, a_{n,j}) < \frac{\varepsilon}{3}.$$

So if we choose $n, m > \max\{N, 3/\varepsilon\}$, and choose $j \in \mathbb{N}$ so that $d(a_{m,j}, x_m) < 1/m$, $d(a_{n,j}, x_n) < 1/n$ and $d(a_{m,j}, a_{n,j}) < \varepsilon/3$, then

$$d(x_m, x_n) \leq d(x_m, a_{m,j}) + d(a_{m,j}, a_{n,j}) + d(a_{n,j}, x_n)$$
$$< \frac{1}{m} + \frac{\varepsilon}{3} + \frac{1}{n} < \frac{\varepsilon}{3} + \frac{\varepsilon}{3} + \frac{\varepsilon}{3} = \varepsilon.$$

Hence (x_m) is a Cauchy sequence in X.

(b) *$d^*([\mathbf{a}]_m, [(x_j)]) \to 0$ as $m \to \infty$.*

Let $\varepsilon > 0$ be given. Then, for all $m \in \mathbb{N}$,

$$d^*([\mathbf{a}]_m, [(x_j)]) \leq d^*([\mathbf{a}]_m, i(x_m)) + d^*(i(x_m), [(x_j)])$$
$$< \frac{1}{m} + \lim_{j \to \infty} d(x_m, x_j).$$

But (x_j) is a Cauchy sequence, and so we can find an $N \in \mathbb{N}$ such that $d(x_m, x_j) < \varepsilon/2$, for all $j, m > N$. Hence if we choose $m > \max\{N, 2/\varepsilon\}$, then

$$d^*([\mathbf{a}]_m, [(x_j)]) < \frac{\varepsilon}{2} + \frac{\varepsilon}{2} = \varepsilon,$$

and so $[\mathbf{a}]_m \to [(x_j)]$ as $m \to \infty$, as required. ∎

Our results are summarized in the following theorem.

> **Theorem 4.6**
>
> Any metric space (X, d) has a completion (X^*, d^*).

We have shown that any metric space can be embedded as a dense subset of some complete metric space. However, the abstract construction that we have given is not usually very helpful in practice, for it does not give us a clear understanding of the completion of any particular space: this must be obtained by special methods. In fact, only a few completions are known in detail, such as $\mathbb{Q}^* = \mathbb{R}$.

Finally, we return to $(C[0, 1], d)$, where d is the integration metric. $\quad d(f, g) = \int_0^1 |f(x) - g(x)|\, dx.$
Theorem 4.6 tells us that there is a complete metric space that contains $(C[0, 1], d)$ as a dense subspace. In fact, the completion of $(C[0, 1], d)$ is the space $L^1[0, 1]$ — the set of *Lebesgue integrable functions* on the interval $[0, 1]$.

Remark

The Lebesgue integral is a generalization of the Riemann integral: it agrees with the Riemann integral for continuous functions, but is also defined for many functions that are not continuous, such as the function $f\colon [0, 1] \to \mathbb{R}$ given by $f(x) = 1$ if x is rational, and 0 otherwise; the Lebesgue integral of f is 0.

Solutions to problems

1.1 (a) If $n, m > N$, then
$$\left|\frac{1}{n} - \frac{1}{m}\right| \leq \frac{1}{n} + \frac{1}{m} < \frac{2}{N}.$$
Now let $\varepsilon > 0$. If $N > 2/\varepsilon$ and $n, m > N$, then
$$\left|\frac{1}{n} - \frac{1}{m}\right| < \frac{2}{N} < \varepsilon.$$
So $(1/n)$ is a Cauchy sequence.

Alternatively, $(1/n)$ is $d^{(1)}$-convergent with limit 0; so, by Lemma 1.1, it is a $d^{(1)}$-Cauchy sequence.

(b) Let $\varepsilon = 1$ and let $N \in \mathbb{N}$. Then
$$|(N+2) - (N+1)| = 1 = \varepsilon$$
and so (n) is not a Cauchy sequence. (We could have used a similar argument for any fixed value of ε.)

1.2 In Problem 2 you saw that the sequence $(1/n)$ is a $d^{(1)}$-Cauchy sequence. Although this sequence is in $(0, 1]$, its limit 0 is not in $(0, 1]$ and so the sequence does not converge in $(0, 1]$. Thus $((0, 1], d^{(1)})$ is incomplete.

2.1 The set $[0, 1]$ is the only closed set, and so it is the only set A for which $(A, d_A^{(1)})$ is complete.

2.2 The continuous image $f(X)$ of a compact space X is compact, by Theorem 2.2 of *Unit C2*. Thus (Y, e) is compact, and so is complete, by Theorem 2.2 of this unit.

If f is not onto, then $f(X)$ is a compact proper subset of Y. Without further information, we cannot determine whether (Y, e) is complete.

2.3 Let (a_n) be a d_0-Cauchy sequence in X, and take $\varepsilon = \frac{1}{2}$. Then there is an $N \in \mathbb{N}$ such that
$$d_0(a_n, a_m) < \varepsilon = \tfrac{1}{2} \quad \text{for all } n, m > N.$$
Since $d_0(a_n, a_m)$ can be only 0 or 1, we deduce that
$$d_0(a_n, a_m) = 0 \quad \text{for all } n, m > N,$$
and so $a_n = a_m$ for all $n, m > N$. Hence the sequence is eventually constant.

(We could have taken ε to be any value in $(0, 1]$.)

2.4 Let
$$r = \min\{d(x_i, x_j) : 1 \leq i, j \leq n, i \neq j\};$$
then $r > 0$, since X is a finite set.

Thus for all $x, y \in X$, either $d(x, y) = 0$, implying that $x = y$, or $d(x, y) \geq r$. Hence if (a_n) is a Cauchy sequence in (X, d), then for $\varepsilon = \frac{1}{2}r$, there is an $N \in \mathbb{N}$ such that
$$d(a_n, a_m) < \varepsilon = \tfrac{1}{2}r \quad \text{for all } n, m > N.$$
In particular,
$$d(a_n, a_{N+1}) < \tfrac{1}{2}r \quad \text{for all } n > N.$$
But this means that $a_n = a_{N+1}$ for all $n > N$, and so the sequence is eventually constant, and hence convergent. Thus (X, d) is complete.

2.5 (a) We must find an isometry from $[0, 1]$ to $[1, 2]$. Define $f : [0, 1] \to [1, 2]$ by $f(x) = x + 1$. Then f is one–one and onto, and it remains only to verify that f preserves distances. So let $x, y \in [0, 1]$; then
$$\begin{aligned} d^{(1)}(f(x), f(y)) &= |(x+1) - (y+1)| \\ &= |x - y| = d^{(1)}(x, y), \end{aligned}$$
and so distances are preserved by f.

Hence f is an isometry and $([0, 1], d^{(1)})$ and $([1, 2], d^{(1)})$ are isometric.

(b) To show that two spaces are not isometric, it is enough to show that there are points in one space that are at a distance not realized in the other space. In this case the distance between the points 0 and 2 in $[0, 2]$ is 2, whereas the distance between any two points in $[0, 1]$ is at most 1. Hence the two spaces are not isometric.

2.6 Suppose that (a_n) is a sequence in X that converges to a. Since f is an isometry,
$$d(f(a_n), f(a)) = d(a_n, a) \to 0 \text{ as } n \to \infty,$$
and so $(f(a_n))$ is a sequence in Y that converges to $f(a)$.

3.1 The point x is a fixed-point of g if and only if
$$g(x) = x,$$
that is, if and only if
$$x - f(x) = x.$$
This is equivalent to $f(x) = 0$. Thus x is a fixed-point of g if and only if x is a zero of f.

3.2 Integrating both sides of the differential equation with respect to s gives
$$\int_0^t \frac{df}{ds}\, ds = \int_0^t \tfrac{1}{2} \cos f(s)\, ds,$$
and so
$$f(t) - f(0) = \int_0^t \tfrac{1}{2} \cos f(s)\, ds \quad \text{for } 0 \leq t \leq 1.$$
Since $f(0) = 0$, this gives the integral equation
$$f(t) = \int_0^t \tfrac{1}{2} \cos f(s)\, ds \quad \text{for } 0 \leq t \leq 1.$$
Thus if we define a mapping T by
$$T(f)(t) = \int_0^t \tfrac{1}{2} \cos f(s)\, ds \quad \text{for } 0 \leq t \leq 1,$$
then any solution of the differential equation is also a fixed-point of this mapping T.

We now show that T is a mapping from $C[0, 1]$ to itself. Since $s \mapsto (\tfrac{1}{2} \cos \circ f)(s)$ is a composite of continuous functions, it is continuous. Also, the mapping F defined by
$$F(g)(t) = \int_0^t g(s)\, ds \quad \text{for } 0 \leq t \leq 1,$$
is a mapping from $C[0, 1]$ to itself (Exercise *A2.9*) and so T is a mapping from $C[0, 1]$ to itself. The

fixed-points of T correspond to the solutions of the integral equation.

3.3 If $X = \emptyset$, then there is nothing to prove. So suppose $X \neq \emptyset$. Let $a \in X$ and let $c = T(a)$. We show that $T(x) = c$, for each $x \in X$.

Let us estimate $d(T(x), T(a))$. Since T has contraction ratio 0,
$$d(T(x), c) = d(T(x), T(a)) = 0 \times d(x, a) = 0,$$
for each $x \in X$. Hence $T(x) = c$ — that is, T is a constant function.

3.4 For $f, g \in C[0,1]$ and $t \in [0,1]$, we have
$$|T(f)(t) - T(g)(t)| = \left| \int_0^t \tfrac{1}{2}(\cos f(s) - \cos g(s)) \, ds \right|$$
$$\leq \tfrac{1}{2} \int_0^t |\cos f(s) - \cos g(s)| \, ds$$
$$\leq \tfrac{1}{2} \int_0^t |f(s) - g(s)| \, ds$$
$$\leq \tfrac{1}{2} \int_0^t d_{\max}(f, g) \, ds$$
$$= \tfrac{1}{2} t \, d_{\max}(f, g) \leq \tfrac{1}{2} d_{\max}(f, g).$$
Hence
$$d_{\max}(T(f), T(g)) \leq \tfrac{1}{2} d_{\max}(f, g),$$
and so T is a contraction mapping with contraction ratio $\tfrac{1}{2}$.

3.5 (a) It is enough to show that T is a Lipschitz mapping on $(0, 1]$ with Lipschitz constant $\tfrac{1}{2}$.

If $x, y \in (0, 1]$, then
$$|T(x) - T(y)| = |\tfrac{1}{2} x - \tfrac{1}{2} y| = \tfrac{1}{2} |x - y|.$$
Hence T is a Lipschitz mapping on $(0, 1]$ with Lipschitz constant $\tfrac{1}{2}$.

(b) If $T(x) = x$, then $\tfrac{1}{2} x = x$ — that is, $x = 0$. So the only possible fixed-point of T is the point 0, which does not belong to $(0, 1]$. Thus T has no fixed-points.

3.6 For all $x, y \in [1, \infty)$,
$$|T(x) - T(y)| = \left| \left(x + \tfrac{1}{x} \right) - \left(y + \tfrac{1}{y} \right) \right|$$
$$= \left| x - y + \tfrac{y - x}{xy} \right|$$
$$= \left| (x - y) \left(1 - \tfrac{1}{xy} \right) \right|$$
$$< |x - y|.$$

If T has a fixed-point x, then $x = T(x)$. This means that $x = x + 1/x$, and so $1/x = 0$, which is impossible. Thus T has no fixed-points.

This does not contradict the Contraction Mapping Theorem, because T is not a contraction mapping with contraction ratio strictly less than 1.

3.7 By Corollary 3.3,
$$|x_T - T^n(0)| \leq \frac{\left(\tfrac{1}{2}\right)^n}{1 - \tfrac{1}{2}} |0 - T^n(0)| = \left(\tfrac{1}{2}\right)^n.$$
This places x_T in an interval of length $(1/2)^{n-1}$. Now
$$((1/2)^{n-1}) = (0, \tfrac{1}{2}, \tfrac{1}{4}, \tfrac{1}{8}, \tfrac{1}{16}, \tfrac{1}{32}, \cdots).$$
The first term in this sequence that is less than 0.1 is the term corresponding to $n = 5$. We have
$$|x_T - T^5(0)| \leq \left(\tfrac{1}{2}\right)^5 = \tfrac{1}{32}.$$
Since $T^5(0) = 0.450$ to 3 d.p. and $\tfrac{1}{32} = 0.03125$,
$$x_T \in [T^5(0) - \tfrac{1}{32}, T^5(0) + \tfrac{1}{32}] \subset [0.41, 0.49].$$
This places x_T in an interval of length less than 0.1.

3.8 We follow the method used in the text immediately before the theorem. Since $0 < m \leq f'(x) \leq M$ on $[a, b]$, we find, for all $x \in [a, b]$,
$$1 - Mr \leq g_r'(x) = 1 - r f'(x) \leq 1 - mr.$$
Hence if $0 < r \leq 1/M$, then $1 - Mr \geq 0$, and so g_r is non-decreasing on $[a, b]$. In this case, for $x \in [a, b]$,
$$g_r(x) \geq g_r(a) = a - r f(a) > a$$
and
$$g_r(x) \leq g_r(b) = b - r f(b) < b.$$
Hence for $0 < r \leq 1/M$, g_r maps $[a, b]$ into itself. Finally, we apply the Mean Value Theorem to g_r to conclude that, if $0 < r \leq 1/M$, then for $x, y \in [a, b]$,
$$|g_r(x) - g_r(y)| \leq \max_{c \in [a, b]} |g_r'(c)| \, |x - y|$$
$$\leq (1 - mr)|x - y|,$$
and so g_r is a contraction mapping on $[a, b]$ with contraction ratio $1 - mr$.

3.9 For all $t \in [0, 1]$,
$$1 + \int_0^t s f(s) \, ds = 1 + \int_0^t s e^{\tfrac{1}{2} s^2} \, ds$$
$$= 1 + \left[e^{\tfrac{1}{2} s^2} \right]_0^t$$
$$= 1 + \left(e^{\tfrac{1}{2} t^2} - 1 \right) = f(t).$$

3.10 By Problem 3.2, any solution to the differential equation must be a fixed-point of the mapping $T : C[0, 1] \to C[0, 1]$, given by
$$T(f)(t) = \int_0^t \tfrac{1}{2} \cos f(s) \, ds \quad \text{for all } 0 \leq t \leq 1.$$
In Problem 3.4, we saw that this is a contraction mapping with contraction ratio $\tfrac{1}{2}$. By the Contraction Mapping Theorem, there is a unique function $f \in C[0, 1]$ that is a fixed-point of T. Since $(s, y) \mapsto \tfrac{1}{2} \cos y$ is a $(d^{(2)}, d^{(1)})$-continuous function, it follows from Theorem 3.5 that f is also the unique solution of the original differential equation. Taking $f_0(s) = 0$ as our initial guess for f, we find that
$$T(f_0) = \tfrac{1}{2} t$$
and
$$T(T(f_0)) = \int_0^t \tfrac{1}{2} \cos \tfrac{1}{2} s \, ds = \sin \tfrac{1}{2} t.$$

We deduce from Corollary 3.3 that
$$d_{\max}(f, \sin \tfrac{1}{2}t) \leq \frac{(\tfrac{1}{2})^2}{1 - \tfrac{1}{2}} d_{\max}(f_0, T(f_0))$$
$$= \frac{\tfrac{1}{4}}{1 - \tfrac{1}{2}} \times \tfrac{1}{2} = \tfrac{1}{4},$$
and so we can take $g(t) = \sin \tfrac{1}{2}t$.

4.1 In *Unit A4*, we saw that $\mathbb{Q}$ is a dense subset of $\mathbb{R}$ with respect to the Euclidean topology, and Theorem 1.6 implies that $(\mathbb{R}, d^{(1)})$ is a complete metric space. Hence we can take $(X^*, d^*) = (\mathbb{R}, d^{(1)})$.

4.2 (a) and (b) are Cauchy sequences of rational numbers, and so are points in X_C.

(c) is not a Cauchy sequence, and so is not in X_C.

4.3 We show that $\sim$ is reflexive, symmetric and transitive on X_C.

Reflexive: $\mathbf{a} \sim \mathbf{a}$
Let $\mathbf{a} = (a_1, a_2, a_3, \ldots) \in X_C$. For each $n \in \mathbb{N}$, $d(a_n, a_n) = 0$, and so $d(a_n, a_n) \to 0$. Hence $\mathbf{a} \sim \mathbf{a}$.

Symmetric: $\mathbf{a} \sim \mathbf{b}$ if and only if $\mathbf{b} \sim \mathbf{a}$
Let $\mathbf{a} = (a_1, a_2, a_3, \ldots)$ and $\mathbf{b} = (b_1, b_2, b_3, \ldots)$ belong to X_C. Since $d(a_n, b_n) = d(b_n, a_n)$ for each $n \in \mathbb{N}$, we deduce that $d(a_n, b_n) \to 0$ if and only if $d(b_n, a_n) \to 0$ as $n \to \infty$. Thus $\mathbf{a} \sim \mathbf{b}$ if and only if $\mathbf{b} \sim \mathbf{a}$.

Transitive: if $\mathbf{a} \sim \mathbf{b}$ and $\mathbf{b} \sim \mathbf{c}$, then $\mathbf{a} \sim \mathbf{c}$
Let $\mathbf{a} = (a_1, a_2, a_3, \ldots)$, $\mathbf{b} = (b_1, b_2, b_3, \ldots)$ and $\mathbf{c} = (c_1, c_2, c_3, \ldots)$ belong to X_C.

If $\mathbf{a} \sim \mathbf{b}$ and $\mathbf{b} \sim \mathbf{c}$, then $d(a_n, b_n) \to 0$ and $d(b_n, c_n) \to 0$ as $n \to \infty$.

The Triangle Inequality now implies that, for each $n \in \mathbb{N}$,
$$0 \leq d(a_n, c_n) \leq d(a_n, b_n) + d(b_n, c_n).$$
And so $d(a_n, c_n) \to 0$ as $n \to \infty$. Hence $\mathbf{a} \sim \mathbf{c}$.

Thus $\sim$ is an equivalence relation on X_C.

4.4 We show that d^* defines a metric on X^*.

In Lemma 4.2 we showed that $d^* : X^* \times X^* \to \mathbb{R}$. We now need to verify properties (M1)–(M3) for a metric space (*Unit A2*, Section 1).

(M1) By definition, $d^*([\mathbf{a}], [\mathbf{b}]) \geq 0$, and
$$d^*([\mathbf{a}], [\mathbf{a}]) = \lim_{n \to \infty} d(a_n, a_n) = 0.$$
If $d^*([\mathbf{a}], [\mathbf{b}]) = 0$, then $d(a_n, b_n) \to 0$; this implies that $\mathbf{a} \sim \mathbf{b}$, and so $[\mathbf{a}] = [\mathbf{b}]$.

(M2) Since d is a metric we have, for $[\mathbf{a}], [\mathbf{b}] \in X^*$,
$$d^*([\mathbf{a}], [\mathbf{b}]) = \lim_{n \to \infty} d(a_n, b_n)$$
$$= \lim_{n \to \infty} d(b_n, a_n) = d^*([\mathbf{b}], [\mathbf{a}]),$$
and so (M2) holds.

(M3) Let $[\mathbf{a}], [\mathbf{b}], [\mathbf{c}] \in X^*$. By the Triangle Inequality,
$$d(a_n, c_n) \leq d(a_n, b_n) + d(b_n, c_n).$$
Thus since Lemma 4.2 implies that $\lim_{n \to \infty} d(a_n, c_n)$, $\lim_{n \to \infty} d(a_n, b_n)$ and $\lim_{n \to \infty} d(b_n, c_n)$ exist,
$$\lim_{n \to \infty} d(a_n, c_n) \leq \lim_{n \to \infty} d(a_n, b_n) + \lim_{n \to \infty} d(b_n, c_n).$$
Hence
$$d^*([\mathbf{a}], [\mathbf{c}]) \leq d^*([\mathbf{a}], [\mathbf{b}]) + d^*([\mathbf{b}], [\mathbf{c}]),$$
as required.

Since (M1)–(M3) are satisfied, d^* is a metric on X^*.

Index

Cauchy sequence, 6
   convergent, 8
complete metric space, 9, 11
   $(C[0,1], d_{\max})$, 13
   $(\mathbb{R}, d^{(1)})$, 9
   $(\mathbb{R}^k, d^{(k)})$, 12
   closed subsets in, 10
completeness, 16
   is a metric invariant, 18
   is not a topological invariant, 16
completion (of a metric space), 31
contraction mapping, 21
Contraction Mapping Theorem, 19, 21, 23, 26
contraction ratio, 21
convergent Cauchy sequence, 8
convergent sequence, 5

dense subspace, 30

fixed-point, 19

incomplete metric space, 9
   $C[0,1]$ with integration metric, 15
integral equation, 20
isometric invariant, 31
isometry, 17

Lebesgue integral, 36
Lipschitz constant, 21
Lipschitz function, 21

metric invariance of completeness, 16
Monotone Convergence Theorem, 5

solving differential equations, 28

uniform convergence, 13
uniformization argument, 14

zeros of real functions, 26